Everest
Trek

Building Math

Integrating Algebra & Engineering

The *Building Math* project was funded by a grant from the GE Foundation, a philanthropic organization of the General Electric Company that works to strengthen educational access, equity, and quality for disadvantaged youth globally.

The interpretations and conclusions contained in this book are those of the authors and do not represent the views of the GE Foundation.

1   2   3   4   5   6   7   8   9   10
ISBN 978-0-8251-6876-5
J. Weston Walch, Publisher
40 Walch Drive
Portland, ME 04103
www.walch.com
Printed in the United States of America

TABLE OF CONTENTS

Introduction . . . . *v*

Organization and Structure of *Everest Trek* . . . . *vii*

Building Math: Pedagogical Approach, Goals, and Methods . . . . *viii*

Enduring Understandings . . . . *ix*

Everest Trek Overview: Story Line and Learning Objectives . . . . *x*

Assessment Opportunities and Materials Lists . . . . *xi*

Everest Trek Master Materials List . . . . *xiii*

Common Core and ITEEA Standards Correlations . . . . *xiv*

Pacing Planning Guide . . . . *xx*

Cup-Stack Team-Building Activity 1

Everest Trek Prerequisite Math Skills 2

Writing Heuristics, or Rules of Thumb 15

Everest Trek Introduction 16

Introducing the Engineering Design Process (EDP) 22

Design Challenge 1: Gearing Up! 28

Design Challenge 2: Crevasse Crisis! 69

Design Challenge 3: Sliding Down! 106

Resources & Appendices 135

    EDP: Engineering Design Process 136

    Math and Engineering Concepts 138

    Important Vocabulary Terms 138

    Rubrics 140

    Student Work Samples 146

Answer Key 155

INTRODUCTION

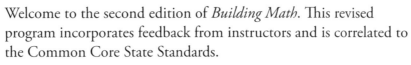

Welcome to the second edition of *Building Math*. This revised program incorporates feedback from instructors and is correlated to the Common Core State Standards.

Building Math is a unique program that integrates real-world math and engineering design concepts with adventurous scenarios that draw students in. The teacher-tested, research-based activities in this program enforce critical thinking skills, teamwork, and problem solving, while bringing students' classroom experiences in line with the Common Core.

Each set of three activities (Design Challenges) forms one unit. The unit's activities are embedded within an engaging fictional situation, providing meaningful contexts for students as they use the engineering design process and mathematical investigations to solve problems. There are three units, and each unit takes about three weeks of class time to implement.

WHAT'S INCLUDED IN THE BUILDING MATH PROGRAM

The instructional materials include reproducible student pages, teacher pages, a DVD of classroom videos for teacher professional development and a Java applet used as a computer model in one of the activities, and a poster showing the engineering design process. The full program is also provided as an interactive PDF on an accompanying CD. This includes all of the content from the book as well as expanded CCSS correlations. The CD is intended to facilitate projecting materials in the classroom and/or printing student pages. You can also print out transparencies from the instructional pages if it suits your needs.

WHY IS ENGINEERING EDUCATION IMPORTANT?

The United States is faced with the challenge of increasing the workforce in quantitative fields (e.g., engineering, science, technology, and math). Schools and teachers play a pivotal role in this challenge. Currently, many students (mostly from underrepresented groups) are not graduating from high school with the necessary math skills to continue studies in college in these quantitative fields. Colleges and industries such as Tufts University and General Electric realize that more active participation in the pre-K–12 system is needed, and have put together this innovative program to help teachers increase math and engineering content in the middle-school curriculum.

(continued)

Algebra and engineering are critical fields that are worth combining. Algebraic reasoning acts as a foundation for higher levels of math learning in secondary and tertiary education, and introducing students to engineering is a way to show them how math is used as a discipline of study and a career path.

BACKGROUND

Building Math was a three-year project funded through the GE Foundation in partnership with the Museum of Science in Boston. One goal of the project was to provide professional development for middle-school teachers in math and engineering, and to explore alternative teaching methods aimed at improving eighth grade students' achievement in algebra and technology. Another project goal was to develop standards-based activities that integrate algebra and engineering using a hands-on, problem-solving, and cooperative-learning approach.

The resulting design challenges were tested by teachers in ten Massachusetts schools that varied in type (public, charter, and independent), location (urban, suburban, rural), and student demographics. Subsequently, hundreds of teachers all over the country used the materials with their students. Their experiences have informed this second edition of *Building Math*, which includes correlations to Common Core State Standards. Further, this edition includes specific, optional suggestions to allow teachers to address additional aspects of the CCSS.

PARTICIPATING RESEARCHERS AND SCHOOLS

Project Investigators: Dr. Peter Y. Wong and Dr. Bárbara M. Brizuela
Project Coordinators: Lori A. Weiss and Wendy Huang
Pilot Schools and Teachers:
- Ferryway School (Malden, MA): Suzanne Collins, Julie Jones
- East Somerville Community School (Somerville, MA): Jack O'Keefe, Mary McClellan, Barbara Vozella
- West Somerville Neighborhood School (Somerville, MA): Colleen Murphy
- Breed Middle School (Lynn, MA): Maurice Twomey, Kathleen White
- Fay School (Southborough, MA): Christopher Hartmann
- The Carroll School (Lincoln, MA): Todd Bearson
- Knox Trail Junior High School (Spencer, MA): Gayle Roach
- Community Charter School of Cambridge (Cambridge, MA): Frances Tee
- Mystic Valley Regional Charter School (Malden, MA): Joseph McMullin
- Pierce Middle School (Milton, MA): Nancy Mikels

ORGANIZATION AND STRUCTURE OF *EVEREST TREK*

Throughout this reproducible book, you will find teacher guide pages followed by one or more student activity pages. Each page is labeled as either a student page or a teacher page.

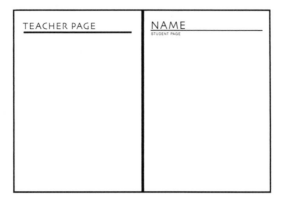

Answers to all activities and discussion questions are found in the answer key at the back of the book.

TIPS, EXTENSIONS, OPTIONAL CCSS ENHANCEMENTS, and **ASSESSMENTS** are labeled in gray boxes.

 INTERESTING INFO is provided in white boxes. This provides additional related information and resources that you may want to share with students.

The following labels are used to indicate whether you will be addressing the whole class, teams, individuals, or pairs:

CLASS **INDIVIDUALS** **TEAMS** **PAIRS**

THE ENGINEERING DESIGN PROCESS

Activities align with the eight-step engineering design process. See page 22 for a lesson plan to introduce the steps.

TABLES AND GRAPHS

Tables and graphs are numbered according to their order of appearance in each design challenge. Those beginning with 1 correspond to Design Challenge 1. Those beginning with 2 correspond to Design Challenge 2. Those beginning with 3 correspond to Design Challenge 3.

BUILDING MATH: PEDAGOGICAL APPROACH, GOALS, AND METHODS

The approach of the *Building Math* program is to engage students in active learning through hands-on, team-based engineering projects that make learning math meaningful to the students (Goldman & Knudson).

The goal of *Building Math* is to encourage the development of conceptual, critical, and creative-thinking processes, as well as social skills, including cooperation, sharing, and negotiation by exhibiting four distinctive methods (Johnson, 1997).

1. **Contextual learning**—Each *Building Math* book involves students in a story line based on real-life situations that pose fictional, but authentic, design challenges. Design contexts invite students to bring ideas, practices, and knowledge from their everyday lives to classroom work. Students apply math skills and knowledge in meaningful ways by using math analysis to connect inquiry-based investigations with creating design solutions.

2. **Peer-based learning**—Students work together throughout the design process. The key to peer-based learning is high amounts of productive, on-task verbalization. By verbalizing their thoughts, students listen to their own and others' thinking. This allows them to evaluate and modify one another's thinking and defend their own ideas. Verbalization also contributes to more precise thinking, especially when teachers use effective questioning techniques to ask students to explain and analyze their and others' reasoning.

3. **Activity-based practice**—*Building Math* uses design challenges (design and construction of a product or process that solves a problem) to focus peer-based learning. Students conduct experiments and systematic investigations; use measuring instruments; carefully observe results; gather, summarize, and display data; build physical models; and analyze costs and trade-offs (Richards, 2005).

4. **Reflective practice**—*Building Math* activities include questions, rubrics, and self-assessment checklists for students to document and reflect on their work throughout each design stage. Teams summarize and present their design solutions to the class, and receive and offer feedback on others' solutions.

REFERENCES

Goldman, S., & Knudsen, J. Learning sciences research and wide-spread school change: Issues from the field. Paper presented at the International Conference of the Learning Sciences (ICLS).

Johnson, S. D. 1997. Learning technological concepts and developing intellectual skills. International Journal of Technology and Design Education, 7, 161–180.

Richards, L. G. 2005. Getting them early: Teaching engineering design in middle-schools. Paper presented at the National Collegiate Inventors & Innovators Alliance (NCIIA). http://www.nciia.org/conf05/cd/supplemental/richards1.pdf

ENDURING UNDERSTANDINGS

In their book, *Understanding by Design,* Wiggins and McTighe advocate that curricula can be built by identifying enduring understandings. An enduring understanding is a big idea that resides at the heart of a discipline, has lasting value outside of the classroom, and requires uncovering of abstract or often misunderstood ideas. The list below details the enduring understandings addressed in *Building Math.* Teachers can consider these to be the ultimate learning goals for the *Building Math* series.

ENGINEERING AND TECHNOLOGICAL LITERACY

1. Technology consists of products, systems, and processes by which humans modify nature to solve problems and meet needs.

2. Design is a creative planning process that leads to other useful products and systems.

3. There is no perfect design.

4. Requirements for design are made up of criteria and constraints.

5. Design involves a set of steps, which can be performed in different sequences and repeated as needed.

6. Successful design solutions are often based on research, which may include systematic experimentation, a trial-and-error process, or transferring existing solutions done by others.

7. Prototypes are working models that can later be improved to become valuable products. Engineers build prototypes to experiment with different solutions for less cost and time than it would take to build full-scale products.

8. Trade-off is a decision process recognizing the need for careful compromises among competing factors.

MATH

1. Math plays a key role in creating technology solutions to meet needs.

2. Mathematical models can represent physical phenomena.

3. Patterns can be represented in different forms using tables, graphs, and symbols.

4. Graphs are useful to visually show the relationship between two variables.

5. Measurement data are approximated values due to tool imprecision and human error.

6. Repeated trials and averages can build one's confidence in measurement data.

7. Mathematical analysis can lead to conclusions to help one make design decisions to successfully meet criteria and constraints.

8. Analyzing data can reveal possible relationships between variables, and support predictions and conjectures.

REFERENCES

International Technology and Engineering Educators Association (ITEEA). 2007. *Standards for Technological Literacy: Content for the Study of Technology.* (Third ed.). Virginia.

Wiggins, G., and J. McTighe. 2005. *Understanding by Design.* (2nd ed.). Prentice Hall.

EVEREST TREK OVERVIEW: STORY LINE AND LEARNING OBJECTIVES

	DESIGN CHALLENGE OVERVIEW	STUDENTS WILL:
DESIGN CHALLENGE 1: GEARING UP!	Students imagine that they are preparing to climb Mount Everest. They learn about the climate conditions and need to design a coat to protect them from the cold and wind. Students research the insulation performance of different materials to help determine which materials to choose for their design. The coat must keep the temperature above 65°F for 30 seconds, not exceed a thickness of 2 cm, and be as low in cost as possible.	• Interpret a line graph. • Locate and represent the range of acceptable values on a graph to meet design criteria. • Extrapolate data based on trends. • Conduct two controlled experiments. • Collect experimental data in a table. • Produce and analyze graphs that relate two variables. • Determine when it's appropriate to use a line graph or a scatter plot to represent data. • Distinguish between independent and dependent variables. • Apply the engineering design process to solve a problem.
DESIGN CHALLENGE 2: CREVASSE CRISIS!	As students are "climbing" Mount Everest, they come to a large crevasse. They study the sagging effect of loading weight onto bridges of different sizes and shapes to design a bridge that will enable them to safely cross the crevasse. The bridge must meet these criteria and constraints: It should support a minimum amount of weight without sagging more than a specific amount, use as few ladders as possible for construction, and be a minimum width to allow for safe crossing.	• Use proportional reasoning to determine dimensions for a scale model. • Use physical and math models. • Conduct two controlled experiments. • Collect experimental data in a table. • Produce and analyze graphs that relate two variables. • Compare rates of change (linear versus non linear relationships). • Distinguish between independent and dependent variables. • Apply the engineering design process to solve a problem.
DESIGN CHALLENGE 3: SLIDING DOWN!	Students have reached the top of Mount Everest. But now they face the problem of altitude sickness and a need to quickly transport their sick classmates down the mountain. Students experiment to find a relationship between the angle of a zip-line and the speed of moving along the zip-line down the mountain. Students use the results of their research to design a zip-line transportation device that meets these criteria and constraints: It must move within a range of acceptable speeds, be stable and secure, and include a return mechanism.	• Conduct a controlled experiment. • Measure angles using a protractor. • Compare and discuss appropriate measures of central tendency (mean, median, mode). • Apply the distance-time-speed formula. • Produce and analyze a graph that relates two variables. • Locate and represent the range of acceptable values on a graph to meet a design criteria. • Distinguish between independent and dependent variables. • Apply the engineering design process to solve a problem.

ASSESSMENT OPPORTUNITIES AND MATERIALS LISTS

The tables on the next two pages list the opportunities for formative assessment by using rubrics, probing students' thinking during class time, and reviewing student responses to certain questions. The tables also show the materials needed for each design challenge. The numbers in the assessment column refer to the steps of the engineering design process.

	Assessment	Materials
DESIGN CHALLENGE 1: GEARING UP!	• **2. Research:** Assess whether students can make a complete graph and correctly represent the experimental data. Use the rubric on page 140. • **2. Research:** Assess whether students can describe the relationship represented in the graph by the variables in the x- and y-axes. • **2. Research:** Assess whether students can extrapolate from the trend of the data. • **2. Research:** Assess whether students can apply the findings of their research to create a rule of thumb to help meet design criteria. • **4. Choose:** Assess engineering drawing based on quality and communication. Use the rubric on page 141. • **6–8. Test, Communicate, Redesign:** Assess written responses and student observations during test, communicate, and redesign steps based on model performance, completeness, and quality of reflection. Use the rubric on page 144. • **Individual Self-Assessment Rubric:** Students can use the checklist on page 66 to determine how well they met behavior and work expectations. • **Team Evaluation:** Students can complete the questions on page 68 to reflect on how well they worked in teams and celebrate successes, as well as make plans to improve teamwork. • **Student Participation Rubric:** Make copies of the rubric on page 145 to score each student's participation in the design challenge.	For each team: • about 5 pieces (15 cm × 15 cm each) of each of the four fabric materials (denim, fleece, nylon, wool) • digital thermometer • 2 frozen ice packs • stopwatch • calculator • 4 different-colored pencils or thin markers • ruler (metric) • chart paper • pack of broad markers (different colors) • transparency markers • transparency For the class: • overhead projector
DESIGN CHALLENGE 2: CREVASSE CRISIS!	• **1. Define:** Assess how students use the scale to find the model measurements given the actual measurements. • **2. Research:** Assess whether students can interpret the graph and how it relates the two variables represented by the x- and y-axes. • **2. Research:** Assess whether students can interpolate data using the graph. • **2. Research:** Assess whether students can apply the findings of their research to create a rule of thumb to help meet design criteria. • **2. Research:** Assess whether students can make a complete graph and correctly represent the experimental data. Use the rubric on page 140. • **4. Choose:** Assess engineering drawing based on quality and communication. Use the rubric on page 142. • **5. Build:** Assess model/prototype artifact based on craftsmanship and completeness. Use the rubric on page 143.	For each team: • 2 identical textbooks, packs of copy paper, or similar equally sized sturdy rectangular prisms • a flat table to work on • 10 pennies • 2 pieces of foam that are each at least 30 cm × 17.5 cm • 1 piece of foam that is at least 30 cm × 12 cm

	Assessment	Materials
DC 2: CREVASSE CRISIS! (CONT.)	• **6–8. Test, Communicate, Redesign:** Assess written responses and student observations during test, communicate, and redesign steps based on model performance, completeness, and quality of reflection. Use the rubric on page 144. • **Individual Self-Assessment Rubric:** Students can use the checklist on page 104 to determine how well they met behavior and work expectations. • **Team Evaluation:** Students can complete the questions on page 105 to reflect on how well they worked in teams and celebrate successes, as well as make plans to improve teamwork. • **Student Participation Rubric:** Make copies of the rubric on page 145 to score each student's participation in the design challenge.	For each team: • graph or white paper • ruler • scissors • tape For the class: • 1 large cup with string attached • 500 pennies
DESIGN CHALLENGE 3: SLIDING DOWN	• **2. Research:** Assess whether students can make a graph that contains all the parts and correctly represents the data using the rubric on page 140. • **2. Research:** Assess whether students can interpret the graph and how it relates the two variables represented by the x- and y-axes. • **2. Research:** Assess whether students can extrapolate data using the graph. • **2. Research:** Assess whether students can apply the findings of their research to create a rule of thumb to help meet design criteria. • **4. Choose:** Assess engineering drawing based on quality and communication using the rubric on page 142. • **5. Build:** Assess model/prototype artifact based on craftsmanship and completeness using the rubric on page 143. • **6-8. Test, Communicate, Redesign:** Assess written responses and student observations during test, communicate, and redesign steps based on model performance, completeness, and quality of reflection using the rubric on page 144. • **Individual Self-Assessment Rubric:** Students can use the checklist on page 133 to determine how well they met behavior and work expectations. • **Team Evaluation:** Students can complete the questions on page 134 to reflect on how well they worked in teams and celebrate successes, as well as make plans to improve teamwork. • **Student Participation Rubric:** Make copies of the rubric on page 145 to score each student's participation in the design challenge.	For each team: • 2 metersticks • piece of fishing line 5 meters long • piece of fishing line 2.05 meters long • piece of straw 5 cm long • protractor • ruler • stopwatch (measures at least to nearest tenth of a second) • chart paper • markers • cardboard (letter-sized) • scissors • tape • calculator • small toy figures

EVEREST TREK MASTER MATERIALS LIST

Qty	Item	C* or R*	1. Gearing Up!	2. Crevasse Crisis!	3. Sliding Down!
PER GROUP					
1	calculator	R	✓		✓
1	digital thermometer (°C)	R	✓		
5	fleece (15 cm × 15 cm)	R	✓		
2	ice packs	R	✓		
2	metersticks	R			✓
1	pack of colored pencils or thin markers (min. 4 colors)	R	✓		
1	pack of markers (broad, different colors)	R	✓		✓
1	pair of scissors	R		✓	✓
10	pennies	R		✓	
5	pieces of denim (15 cm × 15 cm)	R	✓		
5	pieces of nylon (15 cm × 15 cm)	R	✓		
1	protractor	R			✓
1	ruler (metric)	R	✓	✓	✓
2	small toy figures	R			✓
1	stopwatch (or digital timer that shows fractions of seconds)	R	✓		✓
5	wool fabric (15 cm × 15 cm)	R	✓		
8	craft sticks	C		✓	
1	drinking straw (5 cm)	C			✓
1	fishing line (about 9 m)	C			✓
1	piece of cardboard (letter-sized)	C			✓
1	piece of craft foam (30 cm × 12 cm)	C		✓	
2	pieces of craft foam (30 cm × 17.5 cm)	C		✓	
1	roll of tape (invisible)	C		✓	✓
2	sheets of chart paper	C	✓		✓
3	sheets of graph or white paper	C		✓	
PER TEACHER					
1	cup (large) with string attached	R		✓	
500	pennies	R		✓	

*C = Consumable
*R = Reusable

COMMON CORE AND ITEEA STANDARDS CORRELATIONS

The following tables show how each design challenge addresses Common Core State Mathematics Standards and International Technology and Engineering Standards. In the Common Core column, double asterisks (**) denote standards that are not expressly addressed by the design challenges, but that can be addressed by using optional suggestions included in the instructional text for that design challenge. References to the specific pages are included.

	Common Core State Standards for Mathematics (Grades 6–8)[1]	ITEEA Standards for Technological Literacy (STL)[2]
DESIGN CHALLENGE 1: GEARING UP!	**Mathematical Practices** 2. Reason abstractly and quantitatively. 3. Construct viable arguments and critique the reasoning of others. 6. Attend to precision. **Standards** **6.RP.3.** Use ratio and rate reasoning to solve real-world and mathematical problems, e.g., by reasoning about tables of equivalent ratios … double number line diagrams**, or equations**. ***See Optional CCSS Enhancement(s) on page 38.* 　b. Solve unit rate problems including those involving unit pricing and constant speed. **6.NS.3.** Fluently add, subtract, multiply, and divide multi-digit decimals using the standard algorithm for each operation. **6.NS.6.a.** Recognize opposite signs of numbers as indicating locations on opposite sides of 0 on the number line; recognize that the opposite of the opposite of a number is the number itself, e.g., $-(-3) = 3$, and that 0 is its own opposite.** ***See Optional CCSS Enhancement(s) on page 32.* **6.NS.6.b.** Understand signs of numbers in ordered pairs as indicating locations in quadrants of the coordinate plane; recognize that when two ordered pairs differ only by signs, the locations of the points are related by reflections across one or both axes.** ***See Optional CCSS Enhancement(s) on page 41.*	**1F** New products and systems can be developed to solve problems or to help do things that could not be done without the help of technology. **1G** The development of technology is a human activity and is the result of individual and collective needs and the ability to be creative. **1H** Technology is closely linked to creativity, which has resulted in innovation. **2R** Requirements are the parameters placed on the development of a product or system. **2S** Trade-off is a decision process recognizing the need for careful compromises among competing factors. **8E** Design is a creative planning process that leads to useful products and systems. **8F** There is no perfect design. **8G** Requirements for design are made up of criteria and constraints. **9F** Design involves a set of steps, which can be performed in different sequences and repeated as needed. **9G** Brainstorming is a group problem-solving design process in which each person in the group presents his or her ideas in an open forum. **9H** Modeling, testing, evaluating, and modifying are used to transform ideas into practical solutions. **11H** Apply a design process to solve problems in and beyond the laboratory-classroom.

[1]Common Core State Standards. Copyright 2010. National Governor's Association Center for Best Practices and Council of Chief State School Officers. All rights reserved.

[2]International Technology and Engineering Educators Association (ITEEA). 2007. *Standards for Technological Literacy: Content for the Study of Technology.* (Third ed.) Virginia.

Common Core State Standards for Mathematics (Grades 6–8)	ITEEA Standards for Technological Literacy (STL)
6.NS.6.c. Find and position integers and other rational numbers on a horizontal or vertical number line diagram; find and position pairs of integers and other rational numbers on a coordinate plane. **6.EE.8.** Write an inequality of the form $x > c$ or $x < c$ to represent a constraint or condition in a real-world or mathematical problem. Recognize that inequalities of the form $x > c$ or $x < c$ have infinitely many solutions; represent solutions of such inequalities on number line diagrams. **6.EE.9.** Use variables** to represent two quantities in a real-world problem that change in relationship to one another; write an equation to express one quantity, thought of as the dependent variable, in terms of the other quantity, thought of as the independent variable. Analyze the relationship between the dependent and independent variables using graphs and tables, and relate these to the equation. ***See Optional CCSS Enhancement(s) on page 51.* **6.SP.4.** Display numerical data in plots on a number line, including dot plots…. **6.SP.5.** Summarize numerical data sets in relation to their context, such as by: a. Reporting the number of observations. b. Describing the nature of the attribute under investigation, including how it was measured and its units of measurement. **7.RP.2.a.** Decide whether two quantities are in a proportional relationship, e.g., by testing for equivalent ratios in a table or graphing on a coordinate plane and observing whether the graph is a straight line through the origin. **7.RP.2.b.** Identify the constant of proportionality (unit rate) in tables, graphs, … and verbal descriptions of proportional relationships. **7.NS.3.** Solve real-world and mathematical problems involving the four operations with rational numbers.[3]	**11J** Make two-dimensional and three-dimensional representations of the designed solution. **11K** Test and evaluate the design in relation to pre-established requirements, such as criteria and constraints, and refine as needed. **11L** Make a product or system and document the solution.

DESIGN CHALLENGE 1: GEARING UP!

[3]Computations with rational numbers extend the rules for manipulating fractions to complex fractions.

COMMON CORE AND ITEEA STANDARDS CORRELATIONS (*CONTINUED*)

DESIGN CHALLENGE 2: CREVASSE CRISIS!

Common Core State Standards for Mathematics (Grades 6–8)	ITEEA Standards for Technological Literacy (STL)

Mathematical Practices
3. Construct viable arguments and critique the reasoning of others.
5. Use appropriate tools strategically.
6. Attend to precision.
 	
Standards
6.RP.1. Understand the concept of a ratio and use ratio language to describe a ratio relationship between two quantities.
6.RP.3. Use ratio and rate reasoning to solve real-world and mathematical problems, e.g., by reasoning about tables of equivalent ratios … double number line diagrams**, or equations**.
**See Optional CCSS Enhancement(s) on page 73.
  b. Solve unit rate problems including those involving unit pricing and constant speed.
6.NS.3. Fluently add, subtract, multiply, and divide multi-digit decimals using the standard algorithm for each operation.
6.EE.9. Use variables to represent two quantities in a real-world problem that change in relationship to one another; write an equation** to express one quantity, thought of as the dependent variable, in terms of the other quantity, thought of as the independent variable. Analyze the relationship between the dependent and independent variables using graphs and tables….
**See Optional CCSS Enhancement(s) on page 81.
6.SP.4. Display numerical data in plots on a number line, including dot plots….
6.SP.5. Summarize numerical data sets in relation to their context, such as by:
  a. Reporting the number of observations.
  b. Describing the nature of the attribute under investigation, including how it was measured and its units of measurement.
	1F New products and systems can be developed to solve problems or to help do things that could not be done without the help of technology.
	
	1G The development of technology is a human activity and is the result of individual and collective needs and the ability to be creative.
	
	1H Technology is closely linked to creativity, which has resulted in innovation.
	
	2R Requirements are the parameters placed on the development of a product or system.
	
	2S Trade-off is a decision process recognizing the need for careful compromises among competing factors.
	
	8E Design is a creative planning process that leads to useful products and systems.
	
	8F There is no perfect design.
	
	8G Requirements for design are made up of criteria and constraints.
	
	9F Design involves a set of steps, which can be performed in different sequences and repeated as needed.
	
	9G Brainstorming is a group problem-solving design process in which each person in the group presents his or her ideas in an open forum.
	
	9H Modeling, testing, evaluating, and modifying are used to transform ideas into practical solutions.
	
	11H Apply a design process to solve problems in and beyond the laboratory-classroom.

COMMON CORE AND ITEEA STANDARDS CORRELATIONS
(CONTINUED)

	Common Core State Standards for Mathematics (Grades 6–8)	ITEEA Standards for Technological Literacy (STL)
DESIGN CHALLENGE 2: CREVASSE CRISIS!	**7.NS.3.** Solve real-world and mathematical problems involving the four operations with rational numbers. **8.F.5.** Describe qualitatively the functional relationship between two quantities by analyzing a graph (e.g., where the function is increasing or decreasing, linear or nonlinear). Sketch a graph that exhibits the qualitative features of a function that has been described verbally.	**11J** Make two-dimensional and three-dimensional representations of the designed solution. **11K** Test and evaluate the design in relation to pre-established requirements, such as criteria and constraints, and refine as needed. **11L** Make a product or system and document the solution.

COMMON CORE AND ITEEA STANDARDS CORRELATIONS (CONTINUED)

DESIGN CHALLENGE 3: SLIDING DOWN!

Common Core State Standards for Mathematics (Grades 6–8)	ITEEA Standards for Technological Literacy (STL)
Mathematical Practices 2. Reason abstractly and quantitatively. 5. Use appropriate tools strategically. 6. Attend to precision. **Standards** **6.RP.2.** Understand the concept of a unit rate a/b associated with a ratio $a:b$ with $b \neq 0$, and use rate language in the context of a ratio relationship. **6.RP.3.b.** Solve unit rate problems including those involving unit pricing and constant speed. **6.NS.3** Fluently add, subtract, multiply, and divide multi-digit decimals using the standard algorithm for each operation. **6.NS.6.c.** Find and position integers and other rational numbers on a horizontal or vertical number line diagram; find and position pairs of integers and other rational numbers on a coordinate plane. **6.EE.9.** Use variables** to represent two quantities in a real-world problem that change in relationship to one another; write an equation to express one quantity, thought of as the dependent variable, in terms of the other quantity, thought of as the independent variable. Analyze the relationship between the dependent and independent variables using graphs and tables, and relate these to the equation. ***See Optional CCSS Enhancement(s) on page 113.* **6.SP.2.** Understand that a set of data collected to answer a statistical question has a distribution, which can be described by its center, spread**, and overall shape**. ***See Optional CCSS Enhancement(s) on page 113.* **6.SP.4.** Display numerical data in plots on a number line, including dot plots.…	**1F** New products and systems can be developed to solve problems or to help do things that could not be done without the help of technology. **1G** The development of technology is a human activity and is the result of individual and collective needs and the ability to be creative. **1H** Technology is closely linked to creativity, which has resulted in innovation. **2R** Requirements are the parameters placed on the development of a product or system. **2S** Trade-off is a decision process recognizing the need for careful compromises among competing factors. **8E** Design is a creative planning process that leads to other useful products and systems. **8F** There is no perfect design. **8G** Requirements for design are made up of criteria and constraints. **9F** Design involves a set of steps, which can be performed in different sequences and repeated as needed. **9G** Brainstorming is a group problem-solving design process in which each person in the group presents his or her ideas in an open forum. **9H** Modeling, testing, evaluating, and modifying are used to transform ideas into practical solutions.

Common Core State Standards for Mathematics (Grades 6–8)	ITEEA Standards for Technological Literacy (STL)
6.SP.5. Summarize numerical data sets in relation to their context, such as by: a. Reporting the number of observations. b. Describing the nature of the attribute under investigation, including how it was measured and its units of measurement. c. Giving quantitative measures of center (median and/or mean) and variability (interquartile range and/or mean absolute deviation),** as well as describing any overall pattern and any striking deviations from the overall pattern with reference to the context in which the data were gathered. ***See Optional CCSS Enhancement(s) on page 113.* **7.RP.1.** Compute unit rates associated with ratios of fractions, including ratios of lengths, areas and other quantities measured in like or different units. **7.RP.2.a.** Decide whether two quantities are in a proportional relationship, e.g., by testing for equivalent ratios in a table or graphing on a coordinate plane and observing whether the graph is a straight line through the origin.** ***See Optional CCSS Enhancement(s) on page 113.* **7.RP.2.b.** Identify the constant of proportionality (unit rate) in tables, graphs, equations, diagrams, and verbal descriptions of proportional relationships. **7.NS.3.** Solve real-world and mathematical problems involving the four operations with rational numbers. **7.G.2.** Draw (freehand, with ruler and protractor, and with technology) geometric shapes with given conditions. Focus on constructing triangles from three measures of angles or sides, noticing when the conditions determine a unique triangle, more than one triangle, or no triangle.	**11H** Apply a design process to solve problems in and beyond the laboratory-classroom. **11J** Make two-dimensional and three-dimensional representations of the designed solution. **11K** Test and evaluate the design in relation to pre-established requirements, such as criteria and constraints, and refine as needed. **11L** Make a product or system and document the solution. **18F** Transporting people and goods involves a combination of individuals and vehicles. **18G** Transportation vehicles are made up of subsystems, such as structural propulsion, suspension, guidance, control, and support, that must function together for a system to work effectively.

DESIGN CHALLENGE 3: SLIDING DOWN!

PACING PLANNING GUIDE

*ESTIMATED TIME ASSUMES CLASS PERIODS OF 45–50 MINUTES.

Each design challenge takes 5 or more days to complete.

Section name	EDP step[1]	Page number	Estimated time*	Your time estimate
Cup-Stack Team-Building Activity (optional)		1	Day 0	
Everest Trek Prerequisite Math Skills (optional)		2	Day 0	
Everest Trek Introduction		16	Day 1	
Introducing the Engineering Design Process (EDP)		22	Day 1	
DESIGN CHALLENGE 1: GEARING UP!				
Define: Design criteria and constraints are defined.	1	30	Day 1	
Research: Students answer questions about a line graph showing the temperature under a cotton T-shirt when exposed to freezing temperatures.	2	32	Days 1–2	
Research: Students experiment to see how well different materials insulate (single layer); they graph data and answer questions.	2	37	Day 2	
Research: Students experiment to see how well different layers of the same materials insulate; they graph data and answer questions.	2	44	Day 3	
Brainstorm: Given the cost of materials, students individually design a coat to meet criteria and constraints.	3	49	Day 3, homework	
Choose: Team decides on final coat design and draws design.	4	53	Day 4	
Build: Students put together layers of materials for coat prototype.	5	57	Day 4	
Test: Students test prototype to see if criteria are met.	6	59	Day 4	
Communicate: Students answer questions and present to class.	7	61	Days 4–5, Homework	
Redesign: Students answer questions about improving design after hearing about other teams' designs.	8	63	Day 5, homework	

[1]To learn more about the engineering design process (EDP), see pages 136–137.

© Museum of Science (Boston), Wong, Brizuela

Section name	EDP step	Page number	Estimated time	Your time estimate
DESIGN CHALLENGE 2: CREVASSE CRISIS!				
Define: Design criteria and constraints are defined.	1	71	Day 1	
Define: Students solve problems about scale model dimensions.	1	71	Day 1	
Research: Students answer questions and do a short activity comparing the strength of different parts of a craft stick.	2	76	Days 1–2	
Research: Students experiment to see the effect of the width of the ladder on sag when weight is applied; they graph data and answer questions.	2	80	Day 2	
Research: Students experiment to see the effect of the thickness of the ladder on sag when weight is applied; they graph data and answer questions.	2	84	Days 2–3	
Research: Students experiment to see the effect of shape on sag when weight is applied.	2	88	Day 3	
Brainstorm: Students individually sketch ladder-bridge design.	3	90	Days 3–4, homework	
Choose: Team decides on ladder-bridge design and draws design.	4	92	Day 4	
Build: Students build a model of the ladder-bridge.	5	94	Day 4	
Test: Students test the ladder-bridge to see if criteria are met.	6	96	Days 4–5	
Communicate: Students answer questions and share with the whole class.	7	99	Day 5, homework	
Redesign: Students answer questions about improving design after hearing about other teams' designs.	8	101	Days 5–6, homework	

Section name	EDP step	Page number	Estimated time	Your time estimate
DESIGN CHALLENGE 3: SLIDING DOWN!				
Define: Design criteria and constraints are defined.	1	108	Day 1	
Research: Students experiment to see the effect of angle on the time it takes for straw to reach bottom of line; they graph data, answer questions, determine class average data, and calculate speed.	2	110	Days 1–2	
Research: Students answer and discuss question about factors that affect stability.	2	117	Day 2	
Brainstorm: Students individually sketch zip-line design.	3	119	Day 2, homework	
Choose: Team decides on and draws design.	4	121	Day 3	
Build: Students build zip-line model.	5	124	Day 3	
Test: Students test model to see if criteria are met.	6	126	Day 3	
Communicate: Students answer questions and share with class.	7	128	Days 3–4, homework	
Redesign: Students answer questions about improving design after hearing about other teams' designs.	8	130	Day 4, homework	

Teacher Page

CUP-STACK TEAM-BUILDING ACTIVITY *(OPTIONAL)*

Students work and communicate in teams during most of each design challenge in *Everest Trek*. Some pilot teachers found it useful to do some team-building activities prior to the start of the unit. There is a different team-building activity in each of the *Building Math* books.

OBJECTIVES

- Students work together in teams to accomplish a timed task.
- Students practice communication skills.
- Students reflect on their participation in a teamwork setting.

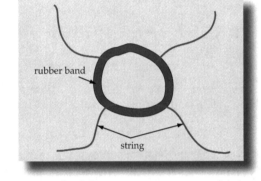

rubber band

string

GROUP SIZE: 3 to 4 students

MATERIALS: You will need a stopwatch or a clock/watch to time 1 minute. Each team needs 15 foam or plastic cups and a rubber band with strings attached (see setup instructions).

SETUP

- Cut string into 60-cm lengths. Tie four strings to the rubber band evenly spaced around the circle. It should look like a sun with four rays coming out.
- Divide the cups into stacks of 15.

PROCEDURES

1. Explain to the class that they will participate in a team-building activity that focuses on accomplishing a task and communication.

2. Distribute a set of materials to each team. Explain that the task is to build a pyramid using the cups within a 1-minute time limit. The pyramid will begin with 4 cups in a row at the base, 3 cups on the next row, then 2 and finally 1 cup at the top. Group members may not touch the cups with their hands or any part of their body, even if the cups fall. Each person may only hold the end of one string attached to the rubber band. Group members must work together to stretch and relax the rubber band to grab each cup and place the cup in the right place.

3. When groups are ready, start timing 1 minute. When 1 minute is up, stop the activity and check each team's progress.

4. Debrief the cup-stack activity with these questions:
    - Was anyone frustrated at all during the activity? If so, how was it handled?
    - Why is teamwork so important for this activity?
    - Did any team come up with a strategy for working together as a team? If so, what was the strategy?
    - Are you ever in a situation where you must use teamwork? Is it always easy for you? Why or why not?
    - What are some skills needed to be good at teamwork?
    - What is so hard about teamwork?
    - How did you contribute to your team? Did you give suggestions? Lead or follow? Encourage or cheer?
    - How would you do the activity differently if you were asked to do it again?

5. Reset and repeat the activity. Give teams a minute to strategize before starting the time. After the task, debrief with these questions:

   - Did your teamwork improve this time? How and why did it improve?
   - Why is good communication important to accomplishing this task?

EVEREST TREK PREREQUISITE MATH SKILLS *(OPTIONAL)*

Students will be using the math skills listed below while doing the Everest Trek design challenges. You may want to review these skills as a short warm-up exercise at the start of the class or as a homework assignment on the night before the challenge when the skill will be used.

STUDENTS SHOULD BE ABLE TO:			
MATH SKILL	1. GEARING UP!	2. CREVASSE CRISIS!	3. SLIDING DOWN!
Interpret a line graph.	✓	✓	✓
Make a line graph.	✓	✓	✓
Use a ruler to measure length in centimeters (to the nearest millimeter).	✓	✓	✓
Round numbers to a given place value (including decimals and powers of 10).	✓	✓	✓
Multiply and add decimals (money) and whole numbers using a calculator or paper/pencil.	✓		
Use a given scale to determine scale model dimensions given actual dimensions.		✓	
Multiply and divide decimals by powers of 10.		✓	
Measure angles with a protractor.			✓
Use the speed = distance/time formula (given distance and time, find speed using a calculator).			✓

* **Bold text indicates key objectives.**

Teacher Page *(continued)*

HOW TO USE THE ACTIVITIES

1. Line Graph Activity (pages 4–8)

    - The goal of this activity is for students to:
        - make a line graph
            - identify independent and dependent variables
            - use convention to put the independent variable on the x-axis and the dependent variable on the y-axis
            - use range of data to set up scales on axes so that the data is well spread out
            - use equal intervals when setting up scales on axes
            - label the axes with data type and unit
            - label the graph with an appropriate title
    - Use Exercise 1 to guide students through the steps of constructing a line graph—particularly steps 2 and 3 (scaling the axes).
    - If students need additional guidance in scaling the axes, use the Line Graph Scaling Examples on pages 9–12.
    - Assign students to do Exercise 2 on their own or with a partner.

2. Using a Scale Activity (pages 13–14)

    - The goal of this activity is for students to:
        - use a scale to determine scale model dimensions given actual dimensions
        - use intuitive proportional reasoning (using drawings, multiplying or dividing by same number to maintain equal ratios)
        - solve proportions as two equal ratios using an equation
    - Go over examples 1 and 2 with the students.
    - Assign students to work through the exercises on their own or with a partner.

LINE GRAPH ACTIVITY *(OPTIONAL)*

A line graph is a way to visually show how two sets of data are related and how they vary depending on each other. Line graphs are particularly effective for showing change over time, predicting what comes next, or estimating what happened in between data points.

HOW TO CONSTRUCT A LINE GRAPH ON PAPER

STEP	WHAT TO DO	HOW TO DO IT
1	Identify the variables.	a. Ask yourself, "Which of the variables did I control or vary?" These values (the independent variables) would be ones that you measure or choose before conducting the experiment. b. The *x*-axis (horizontal) typically represents the independent variable. c. Ask yourself, "Which of the variables was affected as a result of the experiment?" These values (the dependent variables) would be the ones that you measure during the experiment that would correlate one-on-one with the independent variable values. d. The *y*-axis (vertical) typically represents the dependent variable.

2	Determine a scale for each axis.	a. Your goal in determining a scale for the axes is to fit the entire range of data over the available space on the graph paper.
		
		b. Find the range of data (lowest to highest values). If necessary, round down the lowest data value and round up the highest data value to the nearest whole number or power of 10. Find the difference.
		
		c. Count the boxes on the axis that you want to use to represent the range. It is fine to round down. If your data doesn't start at 0, subtract 1 box.
		
		d. Determine the scale: data range to total number of boxes along the axis.
		
		  i. If total number of boxes is greater than the data range,
		    • (total number of boxes) ÷ range = a
		    • a represents the number of boxes in between each whole interval value
		    • round UP if necessary
		  ii. If (total number of boxes) is less than the range,
		    • range ÷ (number of boxes) = b
		    • b represents the interval value of each box
		    • round UP if necessary

3	Number and label each axis.	a. Mark the scale values on each axis. b. Label each axis with the type of data and unit.

4	Plot the data points.	a. Plot each pair of data values (independent-dependent pair) on the graph with a dot.
		b. You may have multiple sets of dependent-variable data. If so, use a different color to plot each set of data pairs.
5	Draw the graph.	Draw a curve or a line that best fits the data points.
6	If necessary, include a key.	If you have multiple sets of dependent-variable data, use a color-coded key to name each set.
7	Title the graph.	a. Your title should include both of the variables that are being compared.
		b. Someone should be able to read the title and know exactly what the graph shows.

EXERCISE 1

The data table below shows the average value of a truck as the mileage on the truck increases. Answer the questions that follow and make a line graph to represent the data.

MILEAGE (KILOMETERS)	0	20,000	40,000	60,000	80,000	100,000	120,000
TRUCK'S VALUE (DOLLARS)	$14,000	$12,000	$8,000	$5,000	$4,000	$3,500	$3,000

1. What is the independent variable? How do you know?
2. What is the dependent variable? How do you know?
3. Write x next to the data table row that contains data values for the x-axis.
4. Write y next to the data table row that contains data values for the y-axis.
5. To scale the x-axis:
   a. Find the range of the data values: _____ to _____.
      The difference is: _____.
   b. Count the number of boxes along the x-axis: _____.
   c. Determine the scale: data range _____ to total of boxes _____.
      i. If the data range is greater than the total number of boxes, calculate (data range) ÷ (total number of boxes) = _____.
      ii. If the data range is less than the total number of boxes, calculate (number of boxes) ÷ (data range) = _____.
   d. Use the above information to label the scale values.
   e. Label the x-axis.

6.   Repeat step 5 to scale the *y*-axis. Show your work below.

7.   Plot the data on the graph.

8.   Give the graph a title that best describes the data shown.

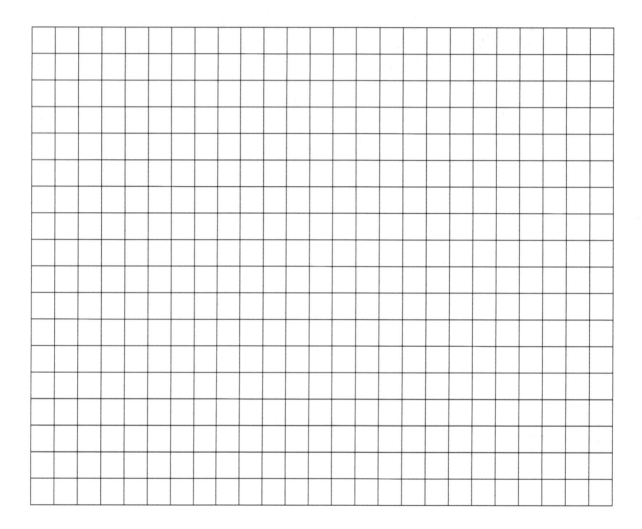

Answer the questions below using the graph.

9.   What was the value of the truck when the mileage was 100,000?

10.   When does the value of the truck decrease the most?

11.   About how much is the truck worth when the mileage is 50,000?

12.   How much will the truck be worth at 140,000 kilometers? How do you know?

EXERCISE 2

The data table below shows the average thickness of annual tree-ring growth in two forests as the trees age. Assume that thicknesses of the tree rings were measured during the same years. A thin ring usually indicates lack of water, forest fires, or a major insect infestation. A thick ring indicates just the opposite. Make a graph of the data.

AGE OF THE TREE (YEARS)	AVERAGE THICKNESS OF THE ANNUAL RINGS IN FOREST A (CM)	AVERAGE THICKNESS OF THE ANNUAL RINGS IN FOREST B (CM)
10	2.0	2.2
20	2.2	2.5
30	3.5	3.6
35	3.0	3.8
50	4.5	4.0
60	4.3	4.5

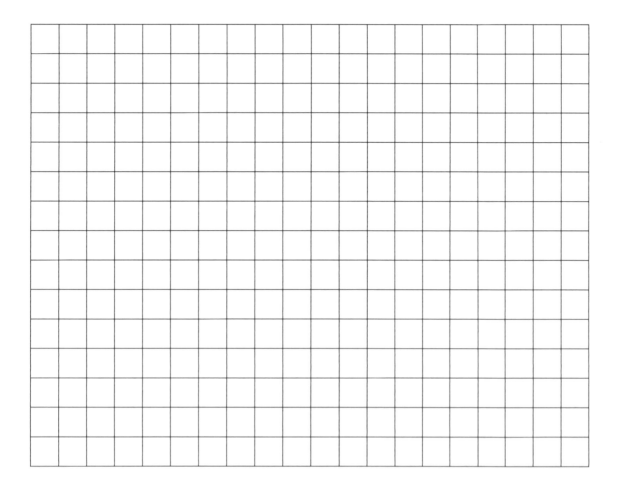

Answer the questions below using the graph.

1. What was the average thickness of annual rings of 20-year-old trees in Forest *B*?

2. How old were the trees in both forests when the average thicknesses of the annual rings were about the same?

3. What would be a reasonable prediction for the average thickness of annual rings of 70-year-old trees in Forest *B*?

4. Based on this data, what can you conclude about Forest *A* and Forest *B*? How do you know? What is your evidence?

Teacher Page

LINE GRAPH SCALING EXAMPLES

Students probably struggle the most with choosing appropriate scales for the axes that fit the full range of data on the available graph paper space. Below are two detailed step-by-step examples of how to scale different kinds of data sets—ones that start with zero and ones that don't, ones that contain decimal values, and ones that contain wide and narrow ranges. Use these to provide direct instruction, as necessary.

SCALING EXAMPLE 1 (DATA STARTS AT 0)

The data table shows the time and distance a car traveled on a road. Make a line graph of the data table.

TIME (HR)	0	1	2	3	4	5	6
DISTANCE (KM)	0	50	100	150	200	230	250

1. The time variable is the independent variable. It'll go on the *x*-axis. The distance variable is the dependent variable. It'll go on the *y*-axis.

2. What is the range of the time data? 0 to 6. The difference is 6.

3. Count the number of boxes on the *x*-axis: 18.

4. You have 18 boxes to represent 6 hours. Since you have more boxes than hours, you'll need multiple boxes to represent each hour. How many boxes would represent 1 hour? Divide the number of boxes by the number of equal intervals: $18 \div 6 = 3$. This means that 3 boxes represent 1 hour.

5. On the *x*-axis, starting at the left-most line and labeling that 0, count 3 boxes and then draw a tick mark and call it 1 hour. Continue counting 3 boxes, drawing tick marks, and labeling each tick mark.

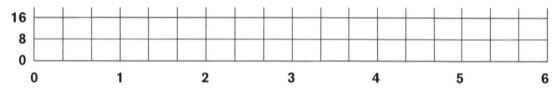

6. What is the range of the distance data? 0 to 250. The difference is 250. Count the number of boxes on the *y*-axis: 32.

7. You have 32 boxes to represent 250 kilometers. Since you have fewer boxes than kilometers, each box needs to represent multiple kilometers. So how many kilometers would 1 box represent? Divide the number of kilometers by the number of boxes: $250 \div 32 = 7.8125$. This means that 1 box would represent 7.8125 kilometers. If you round up to 8 kilometers, 32 boxes would represent more than the necessary 250 range because $32 \times 8 = 256$. But if you round down to 7 kilometers = 1 box, you will only cover a range of $32 \times 7 = 224$, which isn't enough.

    1 box = 8 kilometers

    10 boxes = 80 kilometers

    5 boxes = 40 kilometers

8. On the *y*-axis, starting at the bottom most line and labeling that 0, you can mark the next box 8 (0 + 8 = 8) and continue marking each box 8 more than the previous one. You can also count in increments of 5 boxes and add 40, or count in increments of 10 boxes and add 80. See below. What are the advantages and disadvantages of each kind of labeling?

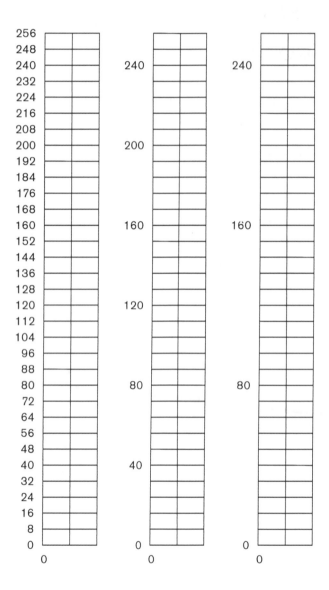

9. Label your axes and give the graph a title.

© Museum of Science (Boston), Wong, Brizuela

SCALING EXAMPLE 2 (DATA DOES NOT START AT 0)

The data table below shows the measurements of thigh lengths of runners and the runners' corresponding times in a 100-meter dash.

THIGH LENGTH (CM)	24	31	37	38	39	42	51	55	62	71
TIME OF 100-METER DASH (SEC)	24.6	25.2	29.9	27.3	22.4	23.0	22.1	25.4	24.6	27.3

1. The independent variable is the thigh length, and the dependent variable is the time of the 100-meter dash. This means that the thigh length values will go on the *x*-axis, and the time of the 100-meter dash will go on the *y*-axis.

2. Range of thigh length is 24 to 71. Round 24 to 20 and 71 to 75 so the approximate range is 20 to 75, for a difference of 55. Why is it okay to round the lowest value down and the highest value up but NOT vice versa? It's okay (even recommended) because your goal in creating this scale is to fit the range of data. Having a range of 55 would definitely fit all your data points, with some room below and above the range. But it's not okay to round the lowest value up and the highest value down because then you won't cover the full range of data values.

3. The number of boxes on the *x*-axis is 23. Round this down to 20 to make it easier to divide. Why is it okay to round down but not okay to round up? It's okay to round down because you don't need to use all the boxes to represent the data. It's not okay to round up because you only have 23 boxes on the axis. If you say you have more, then you will need to draw more boxes to fit the range of data.

    Note: Even if you don't need to round down, subtract at least 1 box from the maximum number of boxes. This is because you're not starting at 0 and would need to start at least 1 box away from the first line.

4. Scale: data range 55 to 20 boxes.

5. Since the data range is higher than the number of boxes, 55 (data range) ÷ 20 (number of boxes) = 2.75. Round this up to 3 so each box represents 3 cm.

6. Since the data doesn't start at 0, draw a squiggly line between the first line and the second line of the *x*-axis to indicate a break in the data values. This means that the first box does not represent the same data interval as the others to follow.

7. Label the second line with the lowest value of your APPROXIMATED range (rounded from 24 to 20). So starting at 20, count 1 box and label the next line whatever the previous value is plus 3. So your scale would be 20, 23, 26, 29, 32, 35, and so forth, with 1 box in between each value. You should have no problem reaching the approximated upper range value of 75 without running out of boxes.

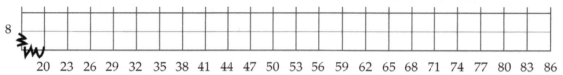

8. Range of 100-meter dash time is 22.1 to 29.9. Round 22.1 down to 22 and 29.9 up to 30 so the approximated range is 22 to 30, for a difference of 8.

9. The number of boxes on the graph is 32.

10. Scale: data range 8 to 32 boxes.

11. Since the number of boxes is higher than the data range, 32 (number of boxes) ÷ 8 (data range) = 4 boxes per 1 second. Furthermore, since the data range includes decimal numbers with digits in the tenth places, each box can represent 0.25 second.

12. Again, since the data doesn't start at 0, draw a squiggly line between the first line and the second line of the y-axis to indicate a break in the data values.

13. Label the second line with the lowest value of the approximated range, which is 22, rounded down from 22.1. Starting from 22, count 4 boxes, mark 23, and so forth. Since there are 4 boxes in between each whole number value, each box represents 0.25 second.

14. Label the axes and give the graph a title.

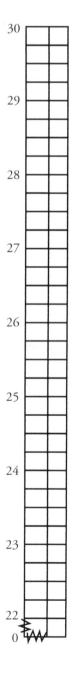

USING A SCALE ACTIVITY

The examples below show two ways to determine scale model dimensions given actual dimensions.

EXAMPLE 1

The scale of a map is _____ cm : 1 m

map: 28 cm

actual: 4 m

This problem can also be solved using a two-step equation.

1. Let x represent the unknown quantity.

2. Write the two equivalent ratios:

$$\frac{\text{map}}{\text{actual}} \quad \frac{28 \text{ cm}}{4 \text{ m}} = \frac{x}{1 \text{ m}}$$

3. Cross-multiply and solve for x:

$$\frac{28 \text{ cm}}{4 \text{ m}} = \frac{x}{1 \text{ m}}$$

$$28 \text{ cm} \cdot 1 \text{ m} = 4 \text{ m} \cdot x$$

$$\frac{28}{4} = \frac{4x}{4}$$

$$7 = x$$

4. Therefore, x represents 7 centimeters.

This problem can also be solved using intuitive proportional reasoning. First, represent the known dimension as a proportion. Since you need to divide 4 m by 4 to get 1 m, divide the numerator (28 cm) by the same 4 in order to keep the fraction equivalent, thus getting 7 centimeters representing 1 meter.

$$\frac{28 \text{ cm} \div 4}{4 \text{ m} \div 4} = \frac{7 \text{ cm}}{1 \text{ m}}$$

EXAMPLE 2

The scale factor for a model is 2.4 cm : 6 m.

model: _____ cm

actual: 2 m

Solve with intuitive proportional reasoning using fraction bars.

6 m			2.4 cm		
÷ into 3 parts to get 2 m per part			÷ into 3 parts to get 0.8 cm per part		
2 m	2 m	2 m	0.8 cm	0.8 cm	0.8 cm

Therefore, an actual length of 2 m is represented by a model length of 0.8 cm.

See Example 1 for other methods of solving the problem.

Name

Use intuitive proportional reasoning or a two-step equation to solve each problem below. Show your work.

1. The scale of a map is ____ cm : 1 m. map: 24 cm actual: 6 m	2. The scale factor for a model is 12 cm : 6 m. model: _____ cm actual: 18 m
3. The scale of a map is 5 cm : 1.5 m. map: 1 cm actual: _____ m	4. The scale factor for a model is 3 cm : 12 m. model: 1 cm actual: _____ m
5. The scale of a map is 0.1 mm : 1 m. map: 1 mm actual: _____ m	6. The scale factor for a model is 8 cm : 1 m. model: _____ cm actual: 0.25 m

WRITING HEURISTICS, OR RULES OF THUMB

Heuristics are rules of thumb people follow in order to make judgments quickly and efficiently.

For each design challenge per class, keep a list on chart paper of research results and other suggestions to consider when making design decisions to meet criteria and constraints. The list should be revisited and added to or refined when students reflect on and discuss the results of their research. Encourage students to use the list when they brainstorm and choose a design. The list should help them find a design that would successfully meet the criteria and constraints.

EXAMPLE

Rules of Thumb for Design Challenge 1: Gearing Up!
Combinations of materials that kept the temperature above 18°C for 30 seconds:
- 2–5 layers of fleece
- 3–5 layers of denim
- 4–5 layers of nylon

Other suggestions to consider when designing:
- Use 1 layer of nylon as the outer layer to keep coat waterproof.
- Using 3 layers of denim costs less and is still effective compared to using 2 layers of fleece.

EVEREST TREK INTRODUCTION

OBJECTIVES
Students will:

- read and understand the story line for the three design challenges in this book

- read background information about climbing Mount Everest and discuss the importance of teamwork

- look at a map of Mount Everest and relate its height to their experiences

1. **CLASS**

    ASK THE CLASS:
    - What is the tallest mountain in the world?
    - Where is it located?

    Possible Answer(s):
    It depends on how you measure the mountain. Mountains are generally measured from sea level. Measuring from sea level, Mount Everest is tallest at 8,848 meters (29,028 feet). Hawaii's Mauna Kea, however, rises 10,203 meters (33,476 feet) from the floor of the Pacific Ocean. Measuring from base to peak, Mauna Kea is the tallest mountain in the world.

2. **CLASS** Read aloud or ask a student to read aloud the introduction.

    ASK THE CLASS:
    - What conditions make climbing Mount Everest so challenging?

    Possible Answer(s):
    Cold temperatures, lack of oxygen, and rough terrain make the climb challenging.
    - What kinds of equipment or gear would a climber need to effectively handle these tough conditions? (Ask for suggestions for each condition.)

    Possible Answer(s):
    - warm clothes for cold temperatures
    - oxygen tank for high altitudes
    - ice picks, ropes, and ladders for climbing and crossing crevices
    - lightweight food and drinks
    - How do you think that climbers can carry all this equipment? Students might not know, but ask them to speculate and pay attention as the introduction continues on page 20.

    Possible Answer(s):
    Local guides called Sherpas help transport equipment and food to drop-off points along the climb.

3. **CLASS** After reading page 20, explain that just as Tenzing Norgay and Edmund Hillary worked together as a team to successfully reach the top of Mount Everest, students will be working in teams to complete the upcoming activities.

ASK THE CLASS:

- What conditions and rules are needed for successful and enjoyable teamwork? What does it look like and sound like?

Possible Answer(s):

Everyone contributes, works hard, takes turns doing tasks, and takes turns leading and following; no one dominates; everyone's ideas are respected and considered; the team stays on task; people smile, are patient and polite, and stay in their seats. People talk quietly, stay on task, and make constructive comments.

Examples include:

- I like your idea because . . .
- I disagree with your because . . .
- I suggest this change because . . .
- How can we make this better?
- Does anyone else have an idea?
- Will you check my work?
- How did you get that answer?
- Let me help you with that.
- You're doing a good job.
- We're a great team.

4. **CLASS** Share with the class that the colorful patterns on the inside of the front cover were inspired by traditional Sherpa designs.

INTERESTING INFO

K2, a mountain also known as Dapsang to local inhabitants, forms part of the Karakoram Range (Himalayas). It lies partly in China and partly on the western side of the Pakistani-administered portion of the Kashmir region. K2 is the world's second highest peak, reaching 8,611 meters (28,251 feet), and is second only to Mount Everest. It was given the name K2 because it was the second peak measured in the Karakoram Range. Italians Achille Compagnoni and Lino Lacedelli were the first climbers to reach the summit of K2 in 1954.

5. **CLASS** Look at the map that shows the climb route for the southern side of Mount Everest.

   ASK THE CLASS:
   - How far is the distance between the base camp and the summit?

   Possible Answer(s): 3,455 meters

   Give some examples to help students relate the distance and the height of the mountain to their experiences so they can appreciate the measurements. For instance, if an adult's height is 1.6 meters, it would take over 5,500 adults this height standing feet-to-head to equal the height of Mount Everest. Or if the school building has four floors, and each floor is about 3 meters high, have students imagine climbing up and down the school stairs about 144 times to equal the distance between the base camp and the summit.

INTERESTING INFO

Mountaineering, also known as mountain climbing, is the sport of attaining or attempting to attain high points in mountainous regions, mainly for the joy of the climb. The pleasures of mountaineering lie not only in achieving the peak, but also in the satisfaction brought about through intense personal efforts.

Mont Blanc was the first great peak ascended (1786). Beginning in the 1930s, a series of successful climbs of mountains in the Himalayas took place. However, the summits of many of the Himalayan Mountains were not reached until the 1950s. The best known of these climbs is the 1953 conquer of Mount Everest by Edmund Hillary and Tenzing Norgay.

Since the 1960s, mountaineering has become an increasingly technical sport, emphasizing the use of specialized anchoring, tethering, and grappling gear in the ascent of vertical rock or ice faces.

EVEREST TREK INTRODUCTION

On the border of Tibet and Nepal, among the beautiful Himalayas, lies the highest mountain in the world, Mount Everest. Often referred to as "The Top of the World," Mount Everest's peak stands at about 8,850 meters above sea level. Imagine standing atop a stack of 5,000 people piled head-to-toe! That is about what it would be like to stand on the summit of Mount Everest. An adventure team from your school has read about some famous mountaineers who have managed to summit this great peak and want to take on the challenge themselves. Using your imagination and developing engineering design skills, you will be joining the adventure team on the trek of a lifetime, battling extreme climate conditions, as you journey to the top of the world!

EVEREST TREK ACTIVITIES

- **INTRODUCING THE ENGINEERING DESIGN PROCESS (EDP)**
  What is engineering? What does an engineer do?

- **DESIGN CHALLENGE 1: GEARING UP!**
  Design a coat to protect your team members from Everest's year-round harsh, frigid weather conditions.

- **DESIGN CHALLENGE 2: CREVASSE CRISIS!**
  Design a lightweight bridge to safely cross a dangerous ice crevasse.

- **DESIGN CHALLENGE 3: SLIDING DOWN!**
  Design an emergency zip-line transportation system to safely and quickly bring your sick teammates down the mountain.

CLIMBING CHALLENGES

Climbing this mountain is no easy feat. You will need to use both physical and mental strength to overcome the treacherous conditions that will confront you on the trek. Frigid temperatures, piercing winds, rough terrain, and the lack of oxygen at high altitudes all pose dangerous threats for climbers. Temperatures on the mountain can dip as low as –60°C. At this temperature, exposed skin can freeze in seconds. Frostbite, which is the freezing of skin, is one of the most common problems for climbers. In addition, the air is thin at high altitudes, so you get less oxygen into your body with each breath. This lack of oxygen to your brain can lead to dizziness, fatigue, and may cause you to hallucinate or even pass out.

SURVIVAL GEAR

With all of nature's threatening forces on the mountain, it is necessary to have specialized mountaineering equipment. You will need personal oxygen tanks, ice-climbing tools, food and water, and much more in order to survive along the trek. You and your teammates are in luck! A group of Sherpas have agreed to be your guide through the trek and will be carrying most of this essential gear. Sherpas are native to the Himalayan region. Because they have lived at high altitudes for many centuries, their bodies have adapted to the thin air on Mount Everest. They are much less prone to altitude sickness. Thus, they are excellent leaders for Everest expeditions.

TEAMWORK

While survival gear is necessary, it may not be the most important thing for a successful climb. Teamwork and the ability to cooperate in a group maybe the most important attribute. In 1953, Tenzing Norgay and Edmund Hillary, a two-man team, conquered Everest. Tied together with 10 meters of rope, these two men greatly relied on each other on their expedition. Although neither man spoke the other's language, they cooperated and stressed the importance of teamwork to reach the summit. Even after they returned from their journey, neither man would admit who actually reached the summit first. Instead, they stated that "it was a partnership" from beginning to end.

MOUNT EVEREST SOUTHERN ROUTE TO SUMMIT

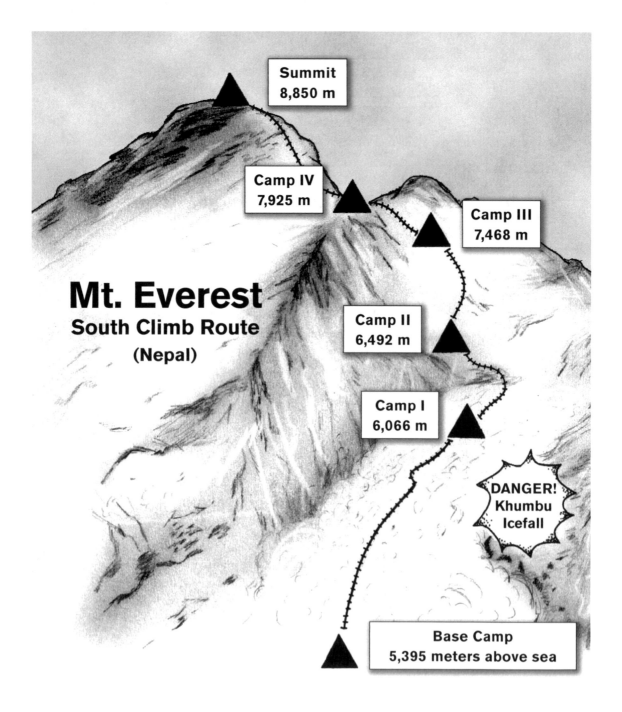

Summit
8,850 m

Camp IV
7,925 m

Camp III
7,468 m

Mt. Everest
South Climb Route
(Nepal)

Camp II
6,492 m

Camp I
6,066 m

DANGER!
Khumbu
Icefall

Base Camp
5,395 meters above sea

INTRODUCING THE ENGINEERING DESIGN PROCESS (EDP)

OBJECTIVE: Students will identify and order the steps of the engineering design process (EDP).

ESTIMATED TIME: 20 minutes

MATERIALS

- 1 set of EDP cards per pair or team of 3 to 4 students

BEFORE YOU TEACH

- Make sets of EDP cards by copying the EDP card templates (pages 24–25) back to back onto card stock, cutting them out, shuffling them, and tying each set with a rubber band.
- Organize students into pairs or teams of 3 to 4 students.

PROCEDURES

1. To get a sense for what students know and think about engineering, ask: "What is an engineer? What does an engineer do?" Students can brainstorm and write their ideas in the space on page 26. If students are struggling to respond to these questions, ask them to list some things that are made by people—for example, houses, roads, cars, televisions, and phones. Explain that engineers have a part in the design and construction of all these things and many more.

2. Explain that all engineers use the engineering design process to help them solve problems in an organized way. Explain that students will use this design process to solve problems in this unit.

3. Distribute one set of EDP cards to each team. Instruct teams to distribute the cards evenly among themselves and take turns reading aloud each step's description. The students' task is to correctly order the steps. Set the time limit to 3 minutes. When debriefing the activity, post each team's steps on the board and compare the lists. Where do teams agree and disagree? Where there is disagreement, ask the teams to explain their rationale for their particular orders. When revealing the "correct" order on pages 136–137, emphasize that the EDP is meant to be a set of guidelines to solve engineering design challenges, but engineers may not always follow all the steps in the same order all the time.

4. Ask the following questions to help students think more about these steps:

    - Why is the step "Communicate" part of the design process? How is it an important step?

      Possible Answer(s): It's important for engineers to communicate their design to other people so they can receive critical feedback and suggestions to improve the design.

    - What do you think happens after the last step, "Redesign"?

      Possible Answer(s): The engineer may go back to an earlier step—which could be as early in the process as "Identify the Problem"—depending on how well the prototype meets specifications. Once the design has gone through several cycles of the design process, it may then be produced on the full-scale level and constructed for real-world use.

5. The EDP matching exercise on page 27 gives students an opportunity to identify the EDP step used in a specific instance of a design challenge.

6. Wrap up the lesson by explaining that students will use these steps to solve three engineering challenges in *Everest Trek* while learning and reinforcing their math skills and understanding. Point out that the octagon on the right-hand corner of each student page shows where students are in the EDP. The description of each EDP step is also on pages 136–137.

DEFINE the problem. What is the problem? What do I want to do? What have others already done? Decide upon a set of specifications (also called "criteria") that your solution should have.

Conduct **RESEARCH** on what can be done to solve the problem. What are the possible solutions? Use the Internet, go to the library, conduct investigations, and talk to experts to explore possible solutions.

BRAINSTORM ideas and be creative! Think about possible solutions in both two and three dimensions. Let your imagination run wild. Talk with your teacher and fellow classmates.

CHOOSE the best solution that meets all the requirements. Any diagrams or sketches will be helpful for later EDP steps. Make a list of all the materials the project will need.

Use your diagrams and list of materials as a guide to **BUILD** a model or prototype of your solution.

TEST and evaluate your prototype. How well does it work? Does it satisfy the engineering criteria?

COMMUNICATE with your fellow peers about your prototype. Why did you choose this design? Does it work as intended? If not, what could be fixed? What were the trade-offs in your design?

Based on information gathered in the testing and communication steps, **REDESIGN** your prototype. Keep in mind what you learned from others in the communication step. Improvements can always be made!

RESEARCH	**DEFINE**
CHOOSE	**BRAINSTORM**
TEST	**BUILD**
REDESIGN	**COMMUNICATE**

INTRODUCING THE ENGINEERING DESIGN PROCESS (EDP)

1. What is engineering? What does an engineer do? Brainstorm and list some of your ideas in the space below.

2. Your team will be given some cards, each naming and describing a step in the engineering design process (EDP). Engineers use the EDP to solve design challenges, just like you will as you go through *Everest Trek*. Your task is to put these steps in a logical order, from Step 1 to Step 8. Be prepared to explain your reasoning for the order you choose.

   STEP 1: _____

   STEP 2: _____

   STEP 3: _____

   STEP 4: _____

   STEP 5: _____

   STEP 6: _____

   STEP 7: _____

   STEP 8: _____

3. Imagine you are part of a team that builds sails and uses them in boat races. Match each sentence to the appropriate step in the engineering design process.

SENTENCE	ENGINEERING DESIGN PROCESS STEP
a. You talk with other sailors to find out how their sails are made.	
b. Your team spends Saturday making a new sail.	
c. After you win the race, you explain the design of the sail to your competitors.	
d. A race is coming up, and your boat needs a new sail. The team decides that the sail must be waterproof, affordable, and strong enough to handle powerful winds.	
e. After meeting to discuss the different designs, your team decides on one design. You find a marina that sells sail material that is strong, waterproof, and cheap.	
f. The sail works pretty well, but when strong gusts of wind blow, the seams rip. Your team resews the seams using stronger stitching.	
g. One week before the race, your team tests the new sail.	
h. Each person on your team sketches a sail design.	

Teacher Page

Design Challenge 1
Gearing Up!
INTRODUCTION

OBJECTIVE: Students will read and understand the problem presented for the first design challenge.

CLASS Together, read the introduction on the next page.

ASK THE CLASS:
- Have you ever been caught outside in cold weather without a coat? Why is it dangerous for the human body to be exposed to freezing temperatures without proper protection?

Possible Answer(s): Exposure can lead to frostbite, organ failure, and so forth.

INTERESTING INFO
Hypothermia results in lowered body temperature and slowed physiological activity. Hypothermia is sometimes artificially induced (usually with ice baths) for certain surgical procedures and cancer treatments. Accidental hypothermia can result from falling into cold water or from overexposure in cold weather. Hypothermia becomes serious when body temperature falls below 35°C (95°F). It is considered an emergency when body temperature goes below 32.2°C (90°F), at which point shivering stops. Pulse, respiration, and blood pressure are depressed. Even when the victim appears dead, revival may be possible with very gradual rewarming.

Design Challenge 1

Gearing Up!

INTRODUCTION

Your team is equipped and ready to climb Mount Everest! Upon arriving at base camp, you turn on the radio and hear the following weather report for the summit of Mount Everest:

"For all you new climbers out there, be prepared for some extreme temperatures on the glacier-covered mountain. In January, the coldest month, the summit temperature averages −36°C and can drop as low as −60°C. In July, the warmest month, the average summit temperature is −19°C. At no time of the year does the temperature on the summit rise above freezing. For the next few weeks, we are expecting an average temperature near −26°C."

Teacher Page

1. DEFINE THE PROBLEM: GEARING UP!

OBJECTIVE: Students will read and understand the criteria and constraints of the design challenge.

CLASS Together, read the engineering criteria and constraints.

ASK THE CLASS:

- What types of insulators have you seen used in real life?
  Possible Answer(s): jackets, blankets, walls
- What materials make up these insulators?
  Possible Answer(s): down feathers, fiberglass, wool
- What characteristics of the material make it an effective insulator?
  Possible Answer(s): thick, fluffy, multi-layered
- Of the four materials you will use in your coat design, which one do you think will be the best insulator? Which one will be the worst? Why?
  Possible Answer(s): Answers will vary.
- Why do you think the criterion is that the temperature must remain above 18˚C?
  Possible Answer(s): 18˚C is the standard room temperature at which most humans are comfortable.

INTERESTING INFO: Math Problem
Base camp instructors are providing climbers with extra clothing because of freezing temperatures in the mountain. The only problem is, your backpack can't exceed 22.75 kg or reaching the summit would become seemingly impossible. Right now, your backpack is 21.75 kg. Which article(s) of clothing will keep you the warmest without exceeding 22.75 kg? Defend your choice.
Possible Answer(s): Basically, any combination that totals less than 1000 g: 1 thermal jacket; 1 parka and 1 T-shirt; 1 fleece jacket and 1 crew shirt; 4 crew shirts

CLOTHING	WEIGHT	SUITABLE TEMPERATURE RANGES
Fleece jacket	618 g	4°C–13°C
Parka	800 g	–7°C–2°C
Thermal jacket	927 g	–18°C– –4°C
Long-sleeve crew shirt	238 g	18°C–24°C
T-shirt	196 g	24°C and up

1. DEFINE THE PROBLEM: GEARING UP!

When any part of your body is exposed to freezing temperatures for more than a few minutes, you are at risk for getting frostbite. Therefore, you will need to dress in outerwear that can keep your body warm in these extremely cold temperatures. Your task is to design a coat that is made of a good insulator to keep you warm, is thin enough to allow you to move easily, and is low in cost.

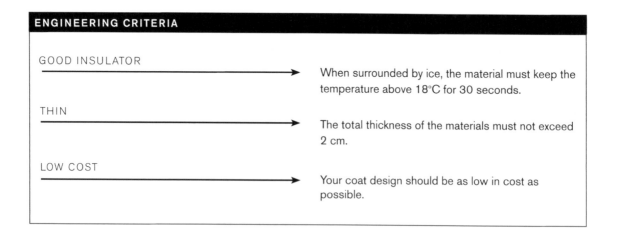

ENGINEERING CRITERIA

GOOD INSULATOR → When surrounded by ice, the material must keep the temperature above 18°C for 30 seconds.

THIN → The total thickness of the materials must not exceed 2 cm.

LOW COST → Your coat design should be as low in cost as possible.

ENGINEERING CONSTRAINTS

You are limited to the following materials for your coat design:

- denim
- fleece
- nylon
- wool

Teacher Page

2. RESEARCH THE PROBLEM: GEARING UP!

RESEARCH PHASE 1: WHAT DO WE KNOW?
OBJECTIVES
Students will:
- interpret a line graph
- connect the research phase to the design challenge

CLASS Together, read Research Phase 1.

ASK THE CLASS:
- Why is the temperature in the lab set at –26°C?
  Possible Answer(s): To simulate the average temperature for the relevant period of time on Mount Everest in the context of this problem
- How can the data from this experiment help someone who is designing a coat for Everest climbers?
  Possible Answer(s): The graph shows how quickly the temperature decreases over time under a cotton T-shirt in a freezing environment. We can see that a cotton T-shirt would not meet the criteria for protecting human bodies. We can conduct similar experiments with other materials to compare their effectiveness as insulators.

Make sure students understand that the temperature of the thermometer starts at 27°C and will decrease over time because the room temperature of the lab is –26°C.

INTERESTING INFO
A common type of thermometer consists of mercury in a glass tube. Calibrated marks on the tube allow the temperature to be read by the length of the mercury within the tube, which varies with temperature. There are many other temperature measuring instruments. In electronics, thermocouples are a widely used temperature-measuring instrument consisting of two wires of different metals joined at each end. Thermocouples are cheap, interchangeable, and can measure a wide range of temperatures. Infrared thermometers measure temperature using electromagnetic radiation emitted from moving objects. By knowing the amount of infrared energy emitted by the object, the object's temperature can be determined.

HELPFUL EXTENSION
Have students calculate and compare slopes between data points, using right triangles, to support their understanding of unit rates and proportions.

OPTIONAL CCSS ENHANCEMENT
To address additional aspects of the Common Core State Standards, emphasize opposites as students construct the axes while graphing Table 1.1.

2. RESEARCH THE PROBLEM: GEARING UP!

RESEARCH PHASE 1: WHAT DO WE KNOW?

Engineers have conducted a study to see how Everest's frigid climate would affect temperature under cotton clothing. Engineers placed a mannequin in a laboratory room at –26°C. A thermometer heated to 27°C was placed inside the mannequin's cotton sleeve. The temperature was recorded every 5 seconds, for 1 minute. The results are provided in the data table and graph below.

Table 1.1: Temperature Under Cotton T-Shirt Over Time

TIME (SECONDS)	0	5	10	15	20	25	30	35	40	45	50	55	60
TEMPERATURE (°C)	26.7	18.5	11.6	5.8	0.9	–3.3	–6.8	–9.8	–12.3	–14.4	–16.3	–17.8	–19.1

Graph 1.1: Temperature Under Cotton T-Shirt Over Time (in –26°C Lab Room)

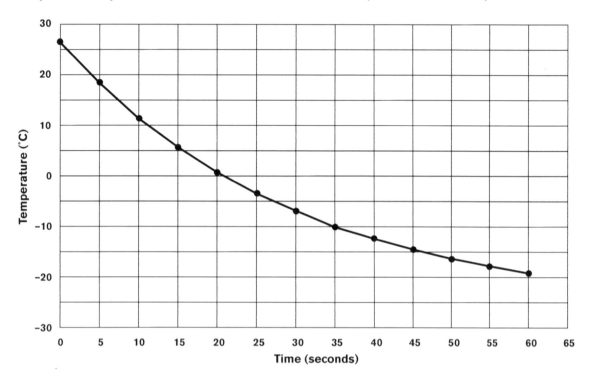

2. RESEARCH THE PROBLEM: GEARING UP!

RESEARCH PHASE 1: WHAT DO WE KNOW? *(CONTINUED)*

OBJECTIVES

Students will:
- locate and represent the range of acceptable values on a graph to meet a design criteria.
- extrapolate data based on trends.
- describe the relationship between two variables.
- describe the rate of change of a line graph.

1. | **PAIRS** | Assign students to work in pairs and use Table 1.1 and Graph 1.1 to answer questions 1–7 in 10 minutes. Things to note (informally assess) as you are circulating the room are listed in the box below.

ASSESSMENT
- Are students able to observe that the rate of change is decreasing, meaning that the temperature decreases more slowly over time?
- Can students extrapolate (predict using the graph's trend) data using the graph?
- Did students realize that the temperature indicated by the graph will eventually approach the lab room temperature of –26°C and not go below that?
- Were students able to represent using a line on the graph how a good insulator would perform?
- Were the questions too easy or too difficult?
- Did students have enough time to complete the question?
- How well are the students working together?
- Were any student responses particularly insightful? Note those for whole-group sharing later.

---

2.   **CLASS**   Go over questions 1–7 using the following discussion points:

- **Questions 1–2:** Discuss whether the answers should have a negative sign. A negative sign, in this case, simply means that the change is negative—the temperature is falling rather than rising.
- **Question 3:** If students are having trouble understanding this, you may want to have students calculate the change in temperature between each 5-second interval over the 60 seconds.
- **Question 4:** Challenge advanced students to find a range of reasonable answers. Some students might use the graph to extend the line to estimate the next value. Others might use second order differences to estimate the next difference. When going over this question, you may want to ask these follow-up questions:
  - Is there more than one correct answer?
  - How do we determine whether an answer is reasonable or not?
  - What would be other reasonable answers? What would be unreasonable answers? Why?

Question 5: Students may mistakenly say that if the temperature decreased by 45.8 degrees in 1 minute, in 60 minutes the temperature would decrease by $45.8 \times 60 = 2{,}748°$. Discuss why this is incorrect. If students have difficulty answering this question correctly, ask them to think about a cup of ice, sitting in a room at room temperature. Over time, the ice would melt and become water at room temperature. Students may benefit from writing inequalities to describe the constraint, or limit, of the temperatures.

Ask the class:

If you left the cup of melted ice out forever, would the temperature keep rising? Would it boil, or would it remain at room temperature?

2. RESEARCH THE PROBLEM: GEARING UP!
RESEARCH PHASE 1: WHAT DO WE KNOW? *(CONTINUED)*

Work with a partner and use Table 1.1 and Graph 1.1 to answer the questions below.

1. How much does the temperature under the cotton clothing change during the first 5 seconds (time = 0 seconds to time = 5 seconds)? _____ ˚C

2. How much does the temperature under the cotton clothing change during the last 5 seconds reported on the graph (time = 55 seconds to time = 60 seconds)? _____ ˚C

3. Compare your answers in questions 1 and 2. Explain what is happening to the temperature as time goes on.

    _____

    _____

4. a. Using the graph, what do you think the temperature under the cotton clothing will be at 65 seconds? _____ ˚C

    b. How did you come up with this prediction?

    _____

    _____

5. a. Imagine that we left the mannequin in the –26 ˚C laboratory room for an entire hour (60 minutes or 3,600 seconds). Predict the temperature under the mannequin's cotton clothing at the end of the hour. _____ ˚C

    b. How did you come up with this prediction?

    _____

    _____

6. A good insulator will be able to keep the temperature of a thermometer in ice above 18 ˚C for 30 seconds. Is the cotton T-shirt a good insulator? Explain why or why not.

    _____

    _____

7. Sketch a line directly on Graph 1.1 showing how the temperature might change if the mannequin was wearing a good insulator. There is more than one correct answer.

2. RESEARCH THE PROBLEM: GEARING UP!
RESEARCH PHASE 2: LEARNING ABOUT MATERIALS
EXPERIMENT 1: HOW WELL DO DIFFERENT MATERAILS INSULATE?

OBJECTIVES

Students will:
- conduct a controlled experiment
- collect experimental data in a table

MATERIALS

For each team:
- 1 piece (15 cm × 15 cm) of each of the 4 fabric materials (denim, fleece, nylon, wool)
- 1 digital thermometer (˚C)
- 2 frozen ice packs
- stopwatch
- 4 different-colored pencils or thin markers

BEFORE YOU TEACH

Put the ice packs in a freezer overnight.

TIP 1

You may use sandwich bags of ice in place of ice packs. Make sure that the bags are tightly sealed.

TIP 2

Transport and store the ice packs in a cooler when not in use.

---

1. **CLASS** Together, read the first paragraph on the next page. Remind students that their coat design must meet the criteria of being able to keep the temperature above 18°C for 30 seconds. Remind students that they have four fabric materials to choose from—denim, fleece, nylon, and wool. Write the names of these four materials on the board.

2. **TEAMS** Give each team 30 seconds to discuss which of the four fabrics they think would best insulate the body and why.

3. **CLASS** Invite each team to share their prediction and mark a tally or an asterisk next to the fabric selected. Together, read the experiment steps and show students the materials they will use.

Emphasize the following:
- Show students how to use the stopwatch, and give them some time to practice using the stopwatch (start, stop, and reset).
- Due to variations in students' hand temperatures, suggest that team members take turns testing how quickly each of them can warm the thermometer to 30°C or higher before deciding who gets the "heater" job.
- When students wrap the material around the bottom of the thermometer stem, it is important that the entire stem is covered (including the tip) but only with one layer of the material. If there are any folds or overlapping pieces, the material will insulate better than it would as a single layer, thus giving students incorrect data.
- Students should follow the experiment's steps carefully and should be sure to leave the thermometer between the ice packs for the entire 30-second trial, collecting all six data values in a single 30-second trial.

4. **TEAMS** Give teams 10 to 15 minutes for this activity. As you circulate, make sure teams are following directions.

OPTIONAL CCSS ENHANCEMENT
To address additional aspects of the Common Core State Standards, direct students to represent the tabular data as a double number line diagram or algebraic equation.

---

2. RESEARCH THE PROBLEM: GEARING UP!

RESEARCH PHASE 2: LEARNING ABOUT MATERIALS

The engineers have shown us that cotton clothing alone will not provide enough insulation to keep you warm in the cold Mount Everest climate. You need to investigate how well other materials insulate.

EXPERIMENT 1: HOW WELL DO DIFFERENT MATERIALS INSULATE?

Student jobs: Assign each member of your team to one of the four jobs listed below.

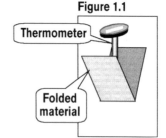

Figure 1.1

Thermometer

Folded material

Heater: _____   Timer: _____

Cooler: _____   Recorder: _____

Step 1: Heater—Clasp your hands around the thermometer stem. Keep your hands there until the temperature is 30°C or higher.

Step 2: Cooler—Fold one layer of material around the thermometer stem (see Figure 1.1). The material should lie flat (no extra folding or overlapping). Make sure that the thermometer stem, including the tip, is surrounded by the material. Watch the thermometer display. When the thermometer reads 27°C, immediately place the folded material and thermometer in between two ice packs and say, "Start timing!"

Step 3: Timer—When you hear the **Cooler** say, "Start timing!" immediately click "start" on your stopwatch. Shout "Time!" at every 5-second interval (when the watch reads 5, 10, 15, 20, 25, and 30 seconds). Leave the thermometer in between the ice packs for the entire 30 seconds.

Step 4: Cooler—Watch the thermometer. Every time the **Timer** shouts "Time!" read the thermometer's temperature aloud. **Recorder**—Record the temperature in Table 1.2 below.

Step 5: After all 30 seconds' worth of data have been recorded for one material, repeat steps 1–4 with a different material.

Table 1.2: Temperature Over Time Inside Different Materials in Ice

MATERIAL	TEMPERATURE (°C) ON THERMOMETER						
	0 sec	5 sec	10 sec	15 sec	20 sec	25 sec	30 sec
denim	27°C						
fleece	27°C						
nylon	27°C						
wool	27°C						

2. RESEARCH THE PROBLEM: GEARING UP!
RESEARCH PHASE 2: LEARNING ABOUT MATERIALS *(CONTINUED)*
EXPERIMENT 1: HOW WELL DO DIFFERENT MATERIALS INSULATE?

OBJECTIVES

Students will:
- produce and analyze a graph that relates two variables
- determine when it is appropriate to use a line graph to represent data
- distinguish between independent and dependent variables

MATERIALS

For each class:
- a sheet of chart paper for the Rules of Thumb list (see page 15)

1. | **CLASS** | If needed, bring the group together and guide them through graphing their data—including labeling the axes and choosing appropriate intervals.

   ASK THE CLASS:
   - What kind of graph should you make to represent your data? Why?

   Possible Answer(s): A line graph may be appropriate because the temperature is changing continuously over time.

   How should you label the axes? Which is the independent variable (*x*-axis)?

   Possible Answer(s): Time, because students controlled that variable.

   Which is the dependent variable (*y*-axis)?

   Possible Answer(s): Temperature, because it varies according to time.

   How do you represent four sets of data on a single graph?

   Possible Answer(s): Use a different color for each line.

2. | **TEAMS** | Instruct teams to graph their data and complete questions 2–6. This is a good time to give students the Rubric for Graphs on page 140. Students can use the rubric as a checklist to make sure that they are producing a high-quality graph. To support students in their graphing, have them graph only *y*-values on a vertical number line. After they have graphed an accurate scale for *y*-values, have them graph *x*- and *y*-coordinates together in a line graph.

Teacher Page *(continued)*

3. **CLASS**

    ASK THE CLASS:
    - Which material seems to make the best insulator? How can you tell?
    - Do everyone's data agree? What might account for differences in data results?
    - When debriefing question 5, you might want to discuss material properties such as water resistance, bulkiness, and other factors that might affect one's choice of insulation materials.
    - Nylon performs the worst as an insulator but could have other useful properties such as being thin, lightweight, waterproof, and wind resistant.
    - Compare the graph from question 1 to the graph from Research Phase 1 on page 33. How are these graphs similar and how are they different?
    - Compare graphs made by different teams of students. Discuss how differences in scale (intervals) on the *x*- and *y*-axes affect the appearance of the data. Invite students to share how they decided on what intervals to use to scale the *x*- and *y*-axes.

4. **CLASS** Introduce heuristics, or rules of thumb (see page 15). Ask students what they learned from Research Phase 2 that they can add to the Rules of Thumb list.

OPTIONAL CCSS ENHANCEMENT

To address additional aspects of the Common Core State Standards, explain quadrants and coordinate pairs as students construct the axes while graphing data from Table 1.2.

---

2. RESEARCH THE PROBLEM: GEARING UP!

RESEARCH PHASE 2: LEARNING ABOUT MATERIALS *(CONTINUED)*
EXPERIMENT 1: HOW WELL DO DIFFERENT MATERIALS INSULATE?

1. Graph your data from Table 1.2 on the grid below. Use a different color for each fabric. Remember to label the axes, color in the key, and give the graph a title. Use the rubric provided by your teacher to assess your work. Then use the graph to answer the questions that follow.

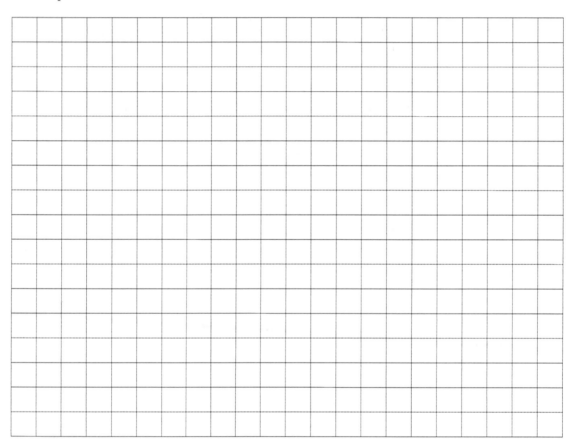

KEY

□ = denim

□ = fleece

□ = nylon

□ = wool

2.   Which of the materials you tested will pass as a good insulator?

_____

3.   Which of the materials seems to be the best insulator?

_____

4.   Why do you think this material is the best insulator?

_____

_____

_____

5.   Do you think there would be any drawbacks, or disadvantages, to using this material? Explain.

_____

_____

_____

6.   Which of the materials seems to be the poorest insulator? Could this material have other properties that could be useful for your coat design?

_____

_____

_____

2. RESEARCH THE PROBLEM: GEARING UP!
RESEARCH PHASE 2: LEARNING ABOUT MATERIALS (CONTINUED)
EXPERIMENT 2: WHAT IS THE EFFECT OF LAYERING ONE MATERIAL?

OBJECTIVES
Students will:
- conduct a controlled experiment
- collect experimental data in a table

MATERIALS
For each team:
- 5 pieces of one fabric material
- digital thermometer (°C)
- 2 frozen ice packs
- stopwatch
- 4 different-colored pencils or thin markers

BEFORE YOU TEACH
Put the ice packs in a freezer overnight.

1. **CLASS**

   ASK THE CLASS:
   - How do you think using more than one layer would affect the insulation ability of the material?
     Possible Answer(s): Multiple layers would increase the ability of the material to insulate from the cold.
   - How might the appearance of the line graph change?
     Possible Answer(s): The temperature would not drop as quickly, and the lines would not be as steep.
   - Which material do you think will perform the best?
     Possible Answer(s): Answers will vary depending on the students' previous experiment results.

2. **TEAMS** Explain to the teams that they will conduct an experiment similar to the first experiment but with different layers of the same material. Assign a different material for each team to test. Afterwards, teams should graph the data and answer the questions on page 48. Students can use the Rubric for Graphs on page 140 to self-assess their graphs. To check that students understand the procedures, ask the following question:
   - How is this experiment different from the first experiment?
     Possible Answer(s): In the first experiment, we used one layer of four different materials. In this experiment, we are using different numbers of layers of the same material. Also, in the first experiment, we recorded the temperature every 5 seconds. In this experiment, we only record the temperature at the end of 30 seconds.

2. RESEARCH THE PROBLEM: GEARING UP!

RESEARCH PHASE 2: LEARNING ABOUT MATERIALS (CONTINUED)
EXPERIMENT 2: WHAT IS THE EFFECT OF LAYERING ONE MATERIAL?

Now that you have investigated how well individual materials insulate, you will take a look at the effect of layering a single material.

Material you will be layering: _____

Student jobs: Assign each member of your group to one of the four jobs listed below.

Heater: _____   Timer: _____

Cooler: _____   Recorder: _____

Step 1: Heater—Clasp your hands around the thermometer stem. Keep your hands there until the temperature is 30°C or higher.

Step 2: Cooler—Fold one layer of material around the thermometer stem (see Figure 1.1 on page 39). The material should lie flat (no extra folding or overlapping). Make sure that the thermometer stem, including the tip, is surrounded by the material. Watch the thermometer display. When the thermometer reads 27°C, immediately place the folded material and thermometer in between two ice packs or sandwich bags of ice and say, "Start timing!"

Step 3: Timer—When you hear the **Cooler** say "Start timing!" immediately click "start" on your stopwatch. Shout "Time!" when the stopwatch reads exactly 30 seconds.

Step 4: Cooler—Watch the thermometer. When the **Timer** shouts "Time!" read the thermometer's temperature aloud. **Recorder**—Record the temperature in Table 1.3 below.

Step 5: Repeat steps 1–4 for the next four layers of your chosen material.

Table 1.3: Number of Layers of Material versus Temperature After 30 Seconds in Ice

NUMBER OF LAYERS	1 LAYER	2 LAYERS	3 LAYERS	4 LAYERS	5 LAYERS
TEMPERATURE AFTER 30 SECONDS (°C)					

2. RESEARCH THE PROBLEM: GEARING UP!

RESEARCH PHASE 2: LEARNING ABOUT MATERIALS *(CONTINUED)*
EXPERIMENT 2: WHAT IS THE EFFECT OF LAYERING ONE MATERIAL?

OBJECTIVES

Students will:
- produce and analyze a graph that represents the relationship between two variables
- determine when it is appropriate to use a scatter plot to represent data
- distinguish between independent and dependent variables

MATERIALS

For each team:
- 1 sheet of transparency with blank graph from page 48
- 1 set of transparency markers

BEFORE YOU TEACH
- Add data table below to each class's Rules of Thumb list.

FABRIC MATERIAL	MIN. NUMBER OF LAYERS NEEDED TO KEEP TEMP ABOVE 18°C FOR 30 SECONDS
denim	
fleece	
nylon	
wool	

After teams complete their graphs, do the following:

1. **CLASS**

    ASK THE CLASS:
    - What were your independent and dependent variables? How did you figure out which is which?
    Possible Answer(s): The independent variable is the number of layers because that is controlled or determined by us. The temperature after 30 seconds is the dependent variable because it changes depending on the number of layers.
    - Is a line graph appropriate for the kind of data? Why or why not?
    Possible Answer(s): A line graph would assume that the number of layers is continuous. Someone could argue that "thickness" is continuous. Also, someone may want to draw a trend line, or connect the dots, in order to see the relationship between the two variables more easily. However, it may be more appropriate to use a scatter plot because there may not be a clear relationship between the two variables.

- Would a bar graph be a good idea here? Why or why not?
  Possible Answer(s): A bar graph might be appropriate because the number of layers could be considered as "discrete" data—meaning that the number of layers are represented by the set of whole numbers (0, 1, 2, 3, . . .) rather than by the entire set of real numbers (that includes fraction parts in between whole numbers). With a bar graph, one can still see trends.

2. | **TEAMS** | Give teams 5 minutes to fix anything they need to on their graphs and transfer their graphs to a transparency. They should also discuss and agree on their answers to questions 3 and 4 and be prepared to share their findings with the class.

3. | **CLASS** | Give each team 2 minutes to share their results by showing the graph on the overhead projector and complete the class data table on the Rules of Thumb list. Ask if students have other suggestions relevant to meeting design criteria and constraints to add to the Rules of Thumb list.

ASSESSMENT
Use the Rubric for Graphs on page 140 to grade each team's graph.

---

2. RESEARCH THE PROBLEM: GEARING UP!

RESEARCH PHASE 2: LEARNING ABOUT MATERIALS *(CONTINUED)*
EXPERIMENT 2: WHAT IS THE EFFECT OF LAYERING ONE MATERIAL?

1. Graph your data from Table 1.3 on the grid below. Remember to label the axes and give the graph a title. Use the rubric provided by your teacher to assess your work. Then use your graph to answer the questions that follow.

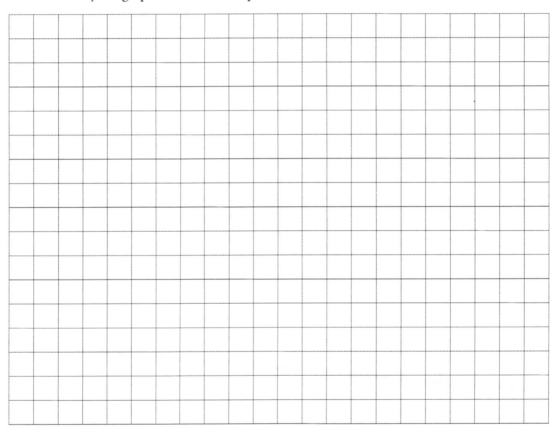

2. Describe the relationship between the number of layers of material and the temperature inside the material after 30 seconds.

    _____

    _____

3. What do you think the temperature would be after 30 seconds for 6 layers? Explain your reasoning.

    _____

    _____

4. What is the minimum number of layers of this material you would need to keep the thermometer above 18°C for 30 seconds? _____ layers

3. BRAINSTORM POSSIBLE SOLUTIONS: GEARING UP!

OBJECTIVE: Students will review the criteria and constraints of the design challenge.

MATERIALS
For each team:
- several pieces of each fabric material
- ruler (metric)

BEFORE YOU TEACH
You may want to give students the option of considering other items for their coat designs, such as a drawstring or snaps. Price these items on a craft store Web site and add them to the table on page 50.

CLASS Instruct students to review the engineering criteria on page 50. Point out that one of the criteria is to make the coat as low in cost as possible. The costs of different materials are listed in Table 1.4 and Table 1.5.

INTERESTING INFO
A zipper is a device used for temporarily joining two edges of fabric together. A zipper mainly consists of two strips of fabric tape, one permanently fixed to each of the two flaps to be joined, and each carrying tens or hundreds of specially shaped metal or plastic teeth. Another part, the slider, operated by hand, rides up and down the two sets of teeth.

The YKK Group is a Japanese group of manufacturing companies. Founded in 1935 by Tadao Yoshida, it was originally named Sanes Shokai. After many name changes, it was finally named YKK in 1994. The name YKK was first registered as a trademark in 1946. Today, the YKK Group is best known for making zippers. The letters YKK appear on many items of clothing.

---

3. BRAINSTORM POSSIBLE SOLUTIONS: GEARING UP!

You've done some great research and are ready to create some possible coat designs. As you brainstorm possible solutions, keep these design requirements in mind.

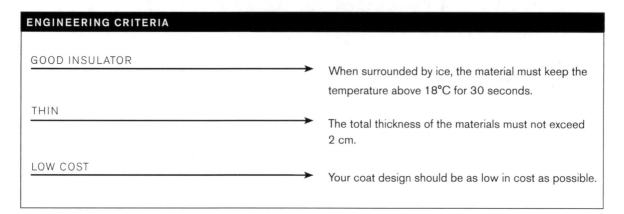

ENGINEERING CRITERIA

GOOD INSULATOR → When surrounded by ice, the material must keep the temperature above 18°C for 30 seconds.

THIN → The total thickness of the materials must not exceed 2 cm.

LOW COST → Your coat design should be as low in cost as possible.

You may combine the four available materials in whatever manner you choose. The cost of one layer of each material to make one coat is listed in Table 1.4 below. Table 1.5 provides the cost of optional items that you may want to include in your coat design.

Table 1.4: Basic Materials for Coat

MATERIAL FOR COAT	COST PER LAYER
denim	$5.25
fleece	$8.50
nylon	$4.75
wool	$10.50

Table 1.5: Optional Items for Coat

OPTIONAL ITEM	COST PER ITEM
zipper	$0.99
plastic button	$0.10
stainless steel button	$0.15
denim pocket	$0.53
fleece pocket	$0.85
nylon pocket	$0.48
wool pocket	$1.05
denim hood	$1.59
fleece hood	$2.55
nylon hood	$1.43
wool hood	$3.15

3. BRAINSTORM POSSIBLE SOLUTIONS: GEARING UP! *(CONTINUED)*

OBJECTIVE: Students will individually brainstorm a coat design.

1. CLASS Together, read and discuss the questions on page 52.
   ASK THE CLASS:
   - Why is it a good idea to individually brainstorm?
     Possible Answer(s): It gives everyone time to think. It allows for greater variety of ideas. It encourages more participation from individuals—especially those who tend to be shy.

2. INDIVIDUALS Instruct students to individually come up with a coat design. The design should include the choice of fabric, the number of layers, as well as any extra items. Students should calculate the cost of their design using the pricing tables. They may have access to the various fabrics to touch and to layer. Remind students to review the Rules of Thumb list to help them make design decisions that meet criteria and constraints.

INTERESTING INFO
Fleece is slowly gaining popularity in the clothing industry. Fleece is lightweight, durable, and perfect for outdoor activities. It was originally made for active use, but now fleece is available in different styles for every possible occasion. It provides warmth without being bulky; it also dries much faster than wool or cotton, and is soft and plush to the touch. In addition, easy care instructions make fleece an ideal clothing material.

OPTIONAL CCSS ENHANCEMENT
To address additional aspects of the Common Core State Standards, direct students to use variables to write an equation when calculating the cost of their coats, using values from the data table.

STUDENT PAGE

BRAINSTORM

3. BRAINSTORM POSSIBLE SOLUTIONS: GEARING UP! *(CONTINUED)*

INDIVIDUAL DESIGN

Using the results of your team's research and the cost information provided in Tables 1.4 and 1.5, think about what might work best for your individual coat design. What materials will you use? How many layers? What optional items would be nice to have? Fill in the charts below as you think about these aspects of your coat design. Calculate the total cost of your idea.

1.   Determine coat layers:

MATERIAL	HOW MANY LAYERS?	× COST PER LAYER	= TOTAL COST FOR EACH MATERIAL
denim		$5.25	
fleece		$8.50	
nylon		$4.75	
wool		$10.50	

Subtotal: _____

2.   Determine extras:

OPTIONAL ITEM FOR YOUR COAT	COST OF OPTIONAL ITEM

Subtotal: _____

3.   What will be the total cost of your idea? _____

Teacher Page

4. CHOOSE THE BEST SOLUTION: GEARING UP!

OBJECTIVES

Students will:

- share their individual brainstorm designs and decide on a team design
- draw to communicate their teams' coat designs

MATERIALS

For each team:
- 1 of each square of fabric—denim, fleece, nylon, and wool
- ruler (metric)

TEAMS Instruct teams to follow the instructions on page 55 to share their individual coat designs and work as a team to use the best ideas to come up with a team design. Students should record their team design on page 55. Once each team agrees on a design, they may proceed to sketch their coat design on page 56. Teams may have access to various fabrics to touch and to layer. Remind students to review the Rules of Thumb list to help them make design decisions that meet the criteria and constraints.

CLASSROOM MANAGEMENT TIPS

- If students need more structure to share their individual coat designs from the Brainstorm step, go over a list of things for each person to share (such as fabrics chosen, total cost, order of layers, how many layers, and optional items); a set time for each person to share (about 1 minute); and behavior expectations for teammates during sharing (e.g., no interruptions, no comments until everyone has shared, active listening).
- Discuss with students how to come to an agreement on a team design. Team members can take turns discussing pros and cons of each design, identify commonalities in designs, compromise on areas of disagreement, and either vote or try to reach consensus.

INTERESTING INFO

Nylon is a thermoplastic silky material that was developed in the 1930s. It was intended to be a synthetic replacement for silk. During World War II, it was used in parachutes, ropes, flak jackets, vehicle tires, and combat uniforms. Today, it is used in fabrics, carpeting, and many other products.

ASSESSMENT

Introduce students to the Rubric for Engineering Drawings on page 141 by showing them examples of student work on pages 146–150, and using the rubric to grade each drawing. You can first assess one drawing with the whole class using the rubric and then have students work in pairs to assess the other drawings. Debrief as a whole class. Students can use the rubric to self-assess their own drawings of the coat design.

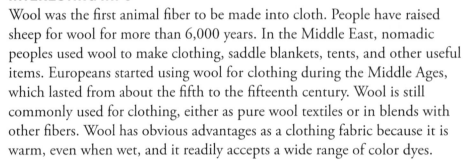

INTERESTING INFO

Wool was the first animal fiber to be made into cloth. People have raised sheep for wool for more than 6,000 years. In the Middle East, nomadic peoples used wool to make clothing, saddle blankets, tents, and other useful items. Europeans started using wool for clothing during the Middle Ages, which lasted from about the fifth to the fifteenth century. Wool is still commonly used for clothing, either as pure wool textiles or in blends with other fibers. Wool has obvious advantages as a clothing fabric because it is warm, even when wet, and it readily accepts a wide range of color dyes.

Denim is a durable woven fabric that is usually colored blue. The name originated in the French Serge de Nîmes. Denim is usually all cotton, though it is sometimes made of a cotton-synthetic blend. Decades of use in the clothing industry, especially in the manufacture of overalls and trousers worn for heavy labor, have demonstrated denim's durability. Denim jeans became extremely popular in the late twentieth century.

CHOOSE

4. CHOOSE THE BEST SOLUTION: GEARING UP!

TEAM DESIGN

Have each member of your team share his or her ideas with the group. For each idea, think about the following:

- What do you like the most about the design?
- Do you think the design will meet the engineering criteria and constraints?

Work as a team to create one "best" coat design. Record the details of your group design in the space below.

1. Determine coat layers:

MATERIAL	HOW MANY LAYERS?	× COST PER LAYER	= TOTAL COST FOR EACH MATERIAL
denim		$5.25	
fleece		$8.50	
nylon		$4.75	
wool		$10.50	

Subtotal: _____

2. Determine extras:

OPTIONAL ITEM FOR YOUR COAT	COST OF OPTIONAL ITEM

Subtotal: _____

3. What will be the total cost of your idea? _____

CHOOSE

4. What will the coat look like? Discuss this with your team and make a sketch of your team's design, based on the materials and optional items you have chosen, in the space below. Then use labels to identify the materials and features of the coat. Show how your design addresses all the criteria and constraints. Be creative! Use the rubric provided by your teacher to check the quality and completeness of your drawing. Practice using the rubric on sample drawings provided by your teacher.

5. BUILD A PROTOTYPE/MODEL: GEARING UP!

OBJECTIVE: Students will put together swatches of fabric according to their designs to represent the prototype.

MATERIALS
For each team:
- enough pieces of fabric materials to build a prototype of their coat design
- ruler (metric)

1. CLASS Together, read page 58, which explains that students will not be building the entire coat but only a prototype that shows the layers.

2. TEAMS Each team should send one member to gather the appropriate materials for their design. The team should then put the layers in the right order (innermost to outermost layers).

INTERESTING INFO
The coat dates as far back as the early Middle Ages. Throughout the centuries, the coat has evolved into many different styles. Coats of the eighteenth and nineteenth centuries include: the duster, a long coat worn by cattlemen to protect clothing from dirt and dust; the basque, a tightly fitted, knee length woman's jacket of the 1870s; and the tailcoat, a late eighteenth century man's coat, preserved in today's white tie and tails. Some more modern coats include the blazer, a nautically-inspired jacket for men or women; the down coat, a warm coat insulated with goose feathers; and the pea coat, a heavy wool double-breasted hip-length jacket.

5. BUILD A PROTOTYPE/MODEL: GEARING UP!

You will not be constructing a model of the entire coat. Instead, you will simply put together the coat's basic layers of materials.

- Look at your design specifications from Design Step 4. Have one team member gather the appropriate number of layers of each material.

- Put the layers together in the order according to your design.

Draw and label a diagram illustrating the order of the layers in your coat design.

6. TEST YOUR SOLUTION: GEARING UP!

OBJECTIVE: Students will follow testing procedures to test their coat prototypes.

MATERIALS

For each team:
- digital thermometer (°C)
- 2 ice packs
- stopwatch
- ruler (metric)

| CLASS | Instruct teams to follow the directions on page 60 to test their designs. Provide the Rubric for Test, Communicate, and Redesign Steps on page 144 so students can assess their work.

ASSESSMENT

Use the Rubric for Test, Communicate, and Redesign Steps on page 144 to assess how well students followed testing procedures.

Name

6. TEST YOUR SOLUTION: GEARING UP!

Use the rubric provided by your teacher to assess your work on the next few pages.

THICKNESS TEST

Your coat may be no more than 2 centimeters thick. Using a ruler, measure the thickness of all of the layers of your coat put together. Is the thickness no more than 2 centimeters?

☐ yes   ☐ no

INSULATOR TEST

Student Jobs: Assign each member of your team to one of the four jobs listed below.

Heater: _____   **Timer:** _____

Cooler: _____   **Recorder:** _____

Step 1: Heater—Clasp your hands around the thermometer stem. Keep your hands there until the temperature is 30°C or higher.

Step 2: Cooler—Fold the layers of material around the thermometer stem once. The material should lie flat (no extra folding or overlapping). Make sure that the thermometer stem, including the tip, is surrounded by the material. Watch the thermometer display. When the thermometer reads 27°C, immediately place the folded material and thermometer in between two ice packs or sandwich bags of ice and say, "Start timing!"

Step 3: Timer—When you hear the **Cooler** say "Start timing!" immediately click "start" on your stopwatch. Shout "Time!" after exactly 30 seconds.

Step 4: Cooler—Watch the thermometer. When the **Timer** shouts "Time!" read the thermometer's temperature aloud. **Recorder**—Record temperature here: _____ °C

After 30 seconds in ice, was the thermometer's temperature above 18°C?   ☐ yes   ☐ no

LOW COST

Record the total cost of your coat here: _____

7. COMMUNICATE YOUR SOLUTION: GEARING UP!

OBJECTIVE: Students will answer questions to reflect on their designs and present their designs to the class.

MATERIALS (OPTIONAL)

For each team:
- chart paper or poster board
- pack of markers

1. **CLASS** After testing, instruct teams to complete questions 1–6. They should be prepared to share their answers with the class. You may want to have each team make a poster to "sell" their design. The poster can include a drawing of the coat, text on how well the design met the criteria, advertising slogans, and the selling price of the coat.

2. **CLASS** As each team presents, you can write notes on a table such as the one below and rank how well each team's design met the criteria to determine one overall "best" design.

TEAM NAME	THICKNESS	INSULATION	COST	PRESENTATION	POSTER

3. **CLASS** Give each team 2 to 3 minutes to share their coat design and test results with the class. Encourage students to ask questions or offer suggestions to improve their classmates' designs.

COMMUNICATE

7. COMMUNICATE YOUR SOLUTION: GEARING UP!

1. Do you think that your coat design was "successful"? Did it meet all of the criteria and constraints? Explain your answer fully.

   _____

   _____

2. Specifically, what are some strengths or advantages of your design? Explain.

   _____

   _____

3. What are some drawbacks or disadvantages of your design? Explain.

   _____

   _____

4. If you could use any materials on Earth, what materials would you use to make your coat?

   _____

   Explain why you would choose these materials. _____

   _____

   _____

5. If you were going to sell your coat, how much would you sell it for? _____
   How would you market your coat? What advertising or slogans would you use to make people want to buy it?

   _____

   _____

6. If the price of your coat is higher than that of the other teams, how would you justify the higher price to your potential customers?

   _____

   _____

Be prepared to share your answers to questions 1–6 with the class.

8. REDESIGN AS NEEDED: GEARING UP!

OBJECTIVE: Students will answer questions to consider how they can redesign their coats.

1. **TEAMS** Instruct teams to use both their own reflections and any suggestions they received from the class to answer questions 1 and 2.

2. **CLASS**

   ASK THE CLASS:

   - Why is a graph useful for representing data?

     Possible Answer(s): A graph can easily show relationships (if any) between two sets of data. It can also show trends—whether upwards or downwards. The steepness of the line can show how quickly or slowly the relationship is changing.

   - How do you determine which variable is dependent and which is independent? Give examples.

     Possible Answer(s): The independent variable is the one that you control—such as the number of layers or the time intervals. The dependent variable changes depending on the independent variable—such as the temperature measured at different time intervals.

   - How do you decide when a line graph is appropriate for representing a set of data?

     Possible Answer(s): A line graph is best suited when the variables contain continuous data—such as time and temperature.

ASSESSMENT
Use the Rubric for Test, Communicate, and Redesign Steps on page 144 to assess students' written work.

---

8. REDESIGN AS NEEDED: GEARING UP!

1. Based on the tests of your coat design, what changes could you make to improve it?

    _____

    _____

    _____

    Explain how these changes would improve your design.

    _____

    _____

    _____

2. Identify one thing that you learned from another group's coat design that you can use to improve your design.

    _____

    _____

    _____

    Explain how this will improve your design.

    _____

    _____

    _____

INDIVIDUAL SELF-ASSESSMENT RUBRIC: GEARING UP!

OBJECTIVE: Students will use a rubric to individually assess their involvement and work in this design challenge.

ASSESSMENT

Assign this reflection exercise as homework. You can write your comments on the lines below the self-assessment and/or use this in conjunction with the Student Participation Rubric on page 145.

STUDENT PAGE

INDIVIDUAL SELF-ASSESSMENT RUBRIC: GEARING UP!

Use this rubric to reflect on how well you met behavior and work expectations during this activity. Check the box next to each expectation that you successfully met.

LEVEL 1	LEVEL 2	LEVEL 3	LEVEL 4	BONUS POINTS
Beginning to meet expectations	Meets some expectations	Meets expectations	Exceeds expectations	
☐ I was willing to work in a group setting.	☐ I met all of the Level 1 requirements.	☐ I met all of the Level 2 requirements.	☐ I met all of the Level 3 requirements.	☐ I helped resolve conflicts on my team.
☐ I was respectful and friendly to my teammates.	☐ I recorded the most essential comments from other group members.	☐ I made sure that my team was on track and doing the tasks for each activity.	☐ I helped my teammates understand the things that they did not understand.	☐ I responded well to criticism.
☐ I listened to my teammates and let them fully voice their opinions.	☐ I read all instructions.	☐ I listened to what my teammates had to say and asked for their opinions throughout the activity.	☐ I was always focused and on task: I didn't need to be reminded to do things; I knew what to do next.	☐ I encouraged everyone on my team to participate.
☐ I made sure we had the materials we needed and knew the tasks that needed to be done.	☐ I wrote down everything that was required for the activity.	☐ I actively gave feedback (by speaking and/or writing) to my team and other teams.	☐ I was able to explain to the class what we learned and did in the activity.	☐ I encouraged my team to persevere when my teammates faced difficulties and wanted to give up.
	☐ I listened to instructions in class and was able to stay on track.	☐ I completed all the assigned homework.		☐ I took advice and recommendations from the teacher about improving team performance and used feedback in team activities.
	☐ I asked questions when I didn't understand something.	☐ I was able to work on my own when the teacher couldn't help me right away.		☐ I worked with my team outside of the classroom to ensure that we could work well in the classroom.
		☐ I completed all the specified tasks for the activity.		

Approximate your level based on the number of checked boxes: _____   Bonus points: _____

Teacher comments: _____

TEAM EVALUATION: GEARING UP!

OBJECTIVE: Students will evaluate and discuss how well they worked in teams.

ASSESSMENT

1. **INDIVIDUALS** Assign this reflection exercise as homework or during quiet classroom time.

2. **TEAMS** Instruct students to share their team evaluation reflections with one another and discuss how they can improve their teamwork during the next activity.

3. **CLASS** Point out any good examples of teamwork and areas to improve during the next activity.

TEAM EVALUATION: GEARING UP!

How well did your team work together to complete the design challenge? Reflect on your teamwork experience by completing this evaluation and sharing your thoughts with your team. Celebrate your successes and discuss how you can improve your teamwork during the next design challenge.

Rate your teamwork. On a scale of 0–3, how well did your team do? 3 is excellent, 0 is very poor. Explain how you came up with that rating.

List things that worked well. Example: We got to our tasks right away and stayed on track.

List things that did not work well. Example: We argued a lot and did not come to a decision that everyone could agree on.

How can you improve teamwork? Make the action steps concrete. Example: We need to learn how to make decisions better. Therefore, I will listen and respond without raising my voice.

Teacher Page

Design Challenge 2
Crevasse Crisis!
INTRODUCTION

OBJECTIVE: Students will read and understand the problem presented for the second design challenge.

CLASS

TELL THE CLASS:

- In the last activity, you designed a coat that would protect you from the cold when climbing Mount Everest. Now imagine that you're climbing the mountain. Look at the map on page 21. What danger is coming up in your climb?
  Possible Answer(s): the Khumbu Icefall

ASK THE CLASS:

- What do you think an "icefall" is? Why might it be dangerous?
  Possible Answer(s): Students might say that it sounds like a weak section of ice that a climber might fall through.

Explain that they will learn more about the icefall and start their next design challenge today. Together, read the introduction on the next page.

ASK THE CLASS:

- What characteristics should these ladders have if you are going to use them to make bridges?
  Possible Answer(s): Ladders should be strong, light, and long.

INTERESTING INFO

A crevasse is formed from fissures or cracks in a glacier resulting from stress produced by movement. A crevasse can range up to 20 meters (65 feet) wide, 45 meters (150 feet) deep, and several hundred meters long. Freshly formed crevasses have clean, straight lines. Over time, crevasses deform and may be hidden when drifting and blowing snow accumulates. Even the most experienced mountaineers can fall into a hidden crevasse. Falling into a hidden crevasse is a mountain climber's worst nightmare. Because of the frequency with which climbers break through the snow over a crevasse and fall in, the crevasse rescue technique is a standard part of climbing education.

Design Challenge 2

Crevasse Crisis!
INTRODUCTION

Your team has started to climb Mount Everest! The coat you designed has kept you warm on your initial travel to Base Camp. As you begin your ascent to Camp I, one of the Sherpa guides warns you that this is one of the most dangerous legs of the trek because you must pass through the Khumbu Icefall. The icefall is full of perilous crevasses, which are deep cracks in the glacial terrain. They are usually not too wide, so it may seem like you could jump across with ease. Famous mountaineer Edmund Hillary tried that once. He managed to leap to the other side, but the fragile terrain could not handle the impact of his landing. Ice crumbled beneath his feet and sent him falling into the crevasse. He would not have made it out alive if it were not for his climbing partner, Tenzing Norgay, who pulled him out by the rope that connected them to each other.

Just as the Sherpa had cautioned, about 1 hour into your climb, you come to a massive crevasse. The shortest path across the crevasse is 2 meters. The depth of the crevasse is 45 meters. Falling into this crevasse would be about the same as falling from the top of the Statue of Liberty! Attempting to jump this crevasse is not worth the risk. Unless you can figure out a way to safely cross this crevasse, your trek will end, and you will have to return to Base Camp, defeated.

Your Sherpa guide informs you that climbers often connect ladders to create bridges across the crevasses. It's worth a shot!

Teacher Page

1. DEFINE THE PROBLEM: CREVASSE CRISIS!

OBJECTIVES

Students will:
- read and understand the criteria and constraints of the design challenge
- use proportional reasoning to determine dimensions for a scale model

1. **CLASS** Together, read page 74 and make sure that students understand the criteria and the constraints.

INTERESTING INFO

A basic crevasse rescue involves two or more climbers who are tied together with a climbing rope, forming a rope team; the standard number of climbers is usually three, with one on each end and one in the middle. When the snow gives way under the victim, the others on the team must immediately prepare for being pulled down. Climbers usually drop down and dig their ice axes and crampons into the snow, facing away from the crevasse if possible, and holding tight until everything stops moving. Many crevasses are small or slanted, and the fallen climber may be able to escape by digging or wiggling out.

Although the mechanical principle of rescue is simple, the reality can be messier: the victim may be injured and/or disoriented from the fall; rescuers on the scene may be anxious or uncertain; equipment and ropes may be scattered everywhere; and everyone will likely already be exhausted and out of breath due to climbing and altitude. There may also be additional crevasses nearby.

Even for experienced climbers, crevasse rescues can be dangerous. Chris Kerrebrock and Jim Wickwire were climbing Mount McKinley when Kerrebrock fell into a crevasse. As the other half of the rope team, Wickwire was pulled down with him. Wickwire was able to get out, but Kerrebrock, who was lodged in upside down, was trapped. Though Wickwire had suffered a dislocated shoulder, he tried unsuccessfully to pull Kerrebrock out. With no one else to assist them, Kerrebrock died of hypothermia and Wickwire had to brave the rest of the mountain alone.

Teacher Page *(continued)*

2. **CLASS** Do question 1 together as a whole group. Use intuitive proportional reasoning or demonstrate how to solve it using proportions (two equal ratios).

   ASK THE CLASS:

   According to this scale, how tall would you be if you were represented in the scale model? (Most students probably don't know how tall they are in meters. Measure a student volunteer and use that student's measurement to calculate.)

   Possible Answer(s): A person who is 1.5 m tall would be represented by a figure 9 cm tall.

3. **TEAMS** Assign students to work in pairs to solve the problems in number 2. For advanced students, encourage them to find another way to solve the problems to verify their answers.

4. **CLASS** Go over answers together.

   - 2a. Besides using a proportion, students can also intuitively solve the problem by reasoning that if 6 cm : 1 m, then 2 m would be twice 6 cm or 12 cm.

   - 2b. Some students might incorrectly reason that since they multiplied by 2 in 2a, they would do the same by multiplying 0.2 m by 2 to get 0.4 cm. Below are three correct ways of reasoning without formally using proportions.

METHOD 1

Use the scale from question 1 (6 cm : 1 m) and divide both sides by 10 to get (0.6 cm : 0.1 m). Then, to find out what 0.2 m represents, multiply 0.6 cm by 2 to get 1.2 cm.

> **6 cm : 1 m**
> ÷ 10 ÷ 10
> **0.6 cm : 0.1 m**
> × 2   × 2
> **1.2 cm : 0.2 m**

METHOD 2

Use the answer from 2a (12 cm : 2 m) and divide both sides by 10 to get 1.2 cm : 0.2 m.

> **12 cm : 2 m**
> ÷ 10 ÷ 10
> **1.2 cm : 0.2 m**

METHOD 3

A third way to visually illustrate the solution is to draw a fraction bar that represents 1 meter. In order to represent 0.2 m, cut the bar into 5 parts, since five 0.2-m pieces make up 1 meter. Draw another bar to represent 6 cm. Since 1 m represents 6 cm in this scale model, divide the 6-cm bar into 5 parts (just like the 1-m bar was divided into 5 parts). Each part of the 6-cm bar represents 1.2 cm, so 0.2 m : 1.2 cm.

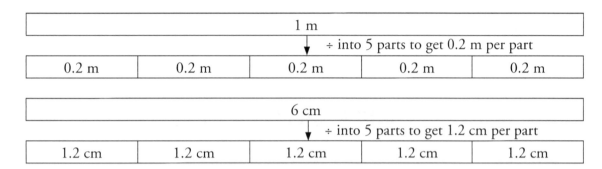

Therefore, 0.2 m represents 1.2 cm.

OPTIONAL CCSS ENHANCEMENT
To address additional aspects of the Common Core State Standards, direct students to represent the scaled data as a double number line diagram or algebraic equation.

1. DEFINE THE PROBLEM: CREVASSE CRISIS!

How can you make a bridge out of flimsy aluminum ladders that will allow you to cross the crevasse without falling in? When there is a significant amount of weight on the ladders, they sag in the middle. If too much weight is placed on them, they will either break or fall to the bottom of the crevasse. Your task is to put ladders together in a way that is strong enough to hold a student, a Sherpa, and a typical load without the ladders sagging too much. Once you design your ladder-bridge, you will radio down to the Sherpas at Base Camp, and they will carry up the necessary number of ladders. The ladders are heavy and bulky, so they are difficult to carry. Therefore, you want to use as few ladders as possible.

ENGINEERING CRITERIA

STRONG ⟶ The ladder-bridge must be able to support an 80-kg Sherpa, a 35-kg student, and an 85-kg load (total load = 200 kg)* without sagging more than 0.2 meters at the center.

MATERIAL USE ⟶ Use as few ladders as possible, and no more than 5 ladders total.

WIDE TOP ⟶ The top of the ladder-bridge must be at least 0.4 meters wide to ensure that you can safely walk across.

* Use 10 pennies to represent the 200 kg load.

ENGINEERING CONSTRAINTS

- Your ladder-bridge design must be long enough to span the 2-meter-wide crevasse.
- For the final design task, you will be provided with the following materials:

ACTUAL MATERIALS	MATERIALS THAT WILL REPRESENT THE ACTUAL MATERIALS
5 aluminum ladders (0.4 m × 5 m)	5 foam strips (2.4 cm × 30 cm)
aluminum shears	scissors
ladder connectors	tape

© Museum of Science (Boston), Wong, Brizuela

FIGURING OUT THE SCALE

1. You will be using 30-cm-long foam strips in your scale model to represent the actual 5-m-long aluminum ladders. If 30 cm in your scale model represents 5 m in real life, then how many centimeters would represent 1 m in real life?

2. Your answer to question 1 gives you the scale you will be using for your model ladder-bridge. Now that you have this scale, you can figure out the engineering criteria for your scale model. Show your work for each of the following questions.

    a. The crevasse is actually 2 meters across. Based on the scale you found in question 1, what distance will your ladder-bridge have to span in your scale model?

    b. The maximum sag allowed for the actual ladder-bridge is 0.2 meters. What is the maximum sag allowed for your scale model?

    c. The actual ladder-bridge must be a minimum of 0.4 meters wide. What is the smallest width you can have for your scale model?

2. RESEARCH THE PROBLEM: CREVASSE CRISIS!
RESEARCH PHASE 1: BENDING UNDER A WEIGHT

OBJECTIVES

Students will:

- brainstorm factors that affect the strength of a bridge and make a conjecture about the effect of thickness on the strength of a wooden craft stick
- test their conjectures about the effect of thickness on a craft stick's strength and write about the results
- connect the results of this research to the engineering design challenge

MATERIALS

For each team:

- 2 craft sticks per person

For each class:

- a sheet of chart paper for the Rules of Thumb list (see page 15)

1. **CLASS** Together, read question 1.

2. **TEAMS** Give teams 1 to 2 minutes to brainstorm at least four other factors that might affect how a bridge sags in the middle.

3. **CLASS** Call on teams to share their ideas.

4. **CLASS** Together, read question 2. Ask students to predict what they think will happen. You might want to hold up a craft stick to demonstrate the different orientations where breakage can occur.

INTERESTING INFO

In engineering mechanics, *bending* describes the behavior of a structural element, usually a beam, when subjected to a lateral load. Bending produces reactive forces inside a beam as the beam attempts to accommodate the flexural load. In cases in which the load is applied downward, the top fibers of the beam are compressed, while the bottom fibers are stretched. Compressive and tensile forces cause stress on the beam. The maximum compressive stress is found at the uppermost edge of the beam, while the maximum tensile stress is located at the lower edge of the beam.

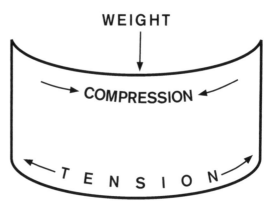

---

5. **TEAMS** Assign students to complete questions 3–5 in about 5 minutes.

6. **CLASS** Debrief the short activity. The stick should be easier to snap on the wide side, because the flat part of the stick is thin and is not very strong. The stick that is oriented to snap on the narrow side is difficult to break because you're trying to break it along the thickest part of the stick.

7. **CLASS** Review the design challenge, criteria, and constraints and ask students what they learned from this research phase that they can add to the Rules of Thumb list.

!

INTERESTING INFO
All engineering materials have mechanical properties that allow for easy classification. The most common properties are modulus of elasticity, tensile strength, and ultimate tensile strength. The modulus of elasticity measures the ability of a material to retain its original shape even after deformation. The tensile strength is a measurement of the stress a material can withstand without permanent deformation. Ultimate tensile strength is the maximum stress a material can withstand. These properties are very important when dealing with material selections for certain applications.

---

2. RESEARCH THE PROBLEM: CREVASSE CRISIS!
RESEARCH PHASE 1: BENDING UNDER A WEIGHT

1. When you drive across a bridge in real life, it sags a little under the weight of the car. The more cars there are on the bridge, the heavier the load the bridge is supporting, and the more the bridge sags.

NO WEIGHT   **A LITTLE WEIGHT**   **A LOT OF WEIGHT**

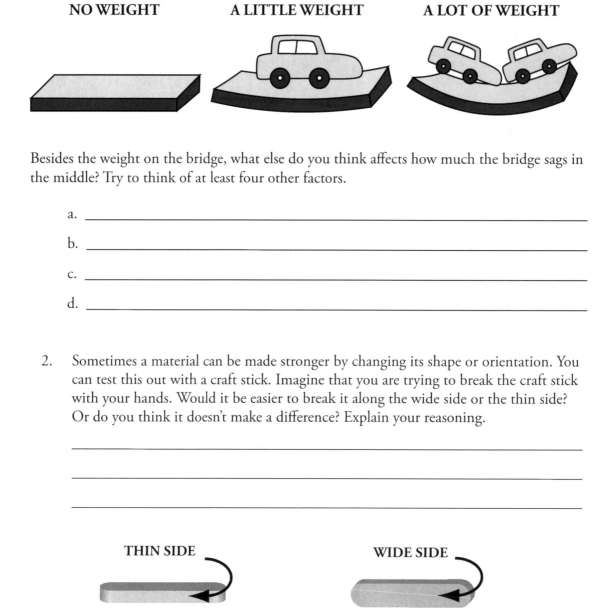

Besides the weight on the bridge, what else do you think affects how much the bridge sags in the middle? Try to think of at least four other factors.

a. _____

b. _____

c. _____

d. _____

2. Sometimes a material can be made stronger by changing its shape or orientation. You can test this out with a craft stick. Imagine that you are trying to break the craft stick with your hands. Would it be easier to break it along the wide side or the thin side? Or do you think it doesn't make a difference? Explain your reasoning.

_____

_____

_____

THIN SIDE   **WIDE SIDE**

3. Hold the craft stick so that you are pressing your thumbs against the wide side, and try to snap the craft stick in half. Could you do it? Was it easy?

    _____

    _____

    _____

4. Get a new craft stick. Hold the craft stick so that you are pressing your thumbs against the thin side, and try to break the craft stick along the thin side. Could you do it? Was it easy?

    _____

    _____

    _____

5. Was your prediction in question 2 correct? Was it easier to break the craft stick along one side than the other? How can you explain what you found?

    _____

    _____

    _____

    _____

    _____

2. RESEARCH THE PROBLEM: CREVASSE CRISIS!
RESEARCH PHASE 2: HOW DOES THE WIDTH AFFECT BENDING?

OBJECTIVES
Students will:
- conduct a controlled experiment
- collect experimental data in a table

MATERIALS
For each team:
- 2 identical textbooks
- a flat table to work on
- 3 pennies
- 1 sheet of graph or white paper
- ruler (metric)
- scissors
- 1 piece of foam that is at least 30 cm × 17.5 cm

BEFORE YOU TEACH
You may precut the "ladders" for your students or assign student helpers to cut ladders for each team before class. For Research Phase 2, each team needs five strips that are each 30 cm long and have the following widths: 1.5 cm, 2.5 cm, 3.5 cm, 4.5 cm, and 5.5 cm.

1. **CLASS** Remind the class of their design challenge—to build a strong bridge that has a minimum width and uses as few "ladders" as possible. To help them come up with a good design, they will research some of the factors that affect strength—width, thickness, and shape. Instruct students to read together and discuss question 1 on page 82.

2. **CLASS** Go over the procedures with the students.
   - **ASK THE CLASS:** Why is it important that the gap between the books is exactly 12 cm? What would happen if the gap changes each time you test a bridge of a different width?

     Possible Answer(s): It is important to keep the gap the same so that the test is "fair," meaning that the test situation is always the same so any differences in results are due to the change in width and not some other change in the test situation.
   - Demonstrate how to mark the center of the 30-cm-long ladder to help make sure that the ladder is centered over the textbooks. There should be 9 cm of the ladder that lies on top of each book.
   - Demonstrate how to properly mark the height of the strip on the graph paper. Students should mark the position of the bottom edge of the ladder (see diagram below).

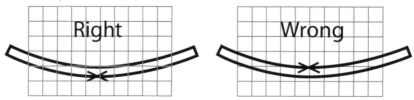

© Museum of Science (Boston), Wong, Brizuela

3. **TEAMS** Instruct teams to conduct the experiment and fill in Table 2.1 on the next page. As you circulate, check to make sure that teams are following the directions and are getting reasonable data.

OBJECTIVES

Students will:
*   produce and analyze a graph that relates two variables
*   distinguish between independent and dependent variables
*   connect the results of this research to the engineering design challenge

MATERIALS

For each class:
*   the Rules of Thumb list

1. **CLASS**

    ASK THE CLASS:

    *   Which is the independent variable and which is the dependent variable? How do you know?

        Possible Answer(s): The width of the ladder is independent because that is what we control—we decide which widths we want to test. The amount of deflection is the dependent variable because it results from changing the width to different values.

    *   Would a line graph be appropriate to graph the data? Why or why not?

        Possible Answer(s): A line graph might be appropriate because the independent variable (width of bridge) could be continuous. You can have width values in between the interval boundaries we selected.

2. **TEAMS** Instruct teams to graph their data, label all parts, and answer questions 3–5 in about 8 minutes. Give students the Rubric for Graphs on page 140 so they can self-assess their graphs.

3. **CLASS** Quickly go over questions 3–5.

4. **CLASS** Review the design challenge, criteria, and constraints and ask students what they learned from this research phase that they can add to the Rules of Thumb list.

OPTIONAL CCSS ENHANCEMENTS

To address additional aspects of the Common Core State Standards, have students compare/contrast their graphs for widths and thicknesses. Discuss which have linear/non-linear qualities. Direct students to write an equation to express their graphs and findings about dependent and independent variables.

STUDENT PAGE

RESEARCH

2. RESEARCH THE PROBLEM: CREVASSE CRISIS!
RESEARCH PHASE 2: HOW DOES THE WIDTH AFFECT BENDING?

1. You need to figure out how to make your ladder-bridge as strong as possible. Someone has suggested you test whether the width of a ladder-bridge affects how strong it is. Do you think that width will affect strength? Explain.

    _____

    _____

    _____

You will investigate how width affects strength with the following experiment:

Step 1: Your team needs to cut out five strips of foam that are each exactly 30 cm long. The widths of the strips should be:
a. 1.5 cm   b. 2.5 cm   c. 3.5 cm
d. 4.5 cm   e. 5.5 cm

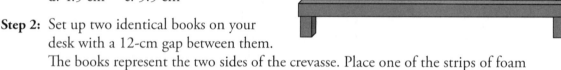

Step 2: Set up two identical books on your desk with a 12-cm gap between them.
The books represent the two sides of the crevasse. Place one of the strips of foam across the gap. Each strip of foam represents a ladder.

Step 3: Hold a sheet of graph paper behind the ladder. Make a mark to indicate the height of the bottom side of the ladder without any weight on it. Hold a book behind the graph paper to provide a hard surface to make the mark.

Step 4: Carefully add 1 penny to the top of the ladder, right in the middle. Hold the sheet of graph paper up again and mark the new height of the bottom side of the ladder at its lowest point. Measure the distance between your two marks. This is the amount of deflection, or the amount that your ladder sags under the weight of 1 penny. Record this value in Table 2.1 below.

Step 5: Repeat steps 2–4 for the other four ladder widths.

Table 2.1: Width versus Deflection of a Ladder Supporting 1 Penny

WIDTH OF LADDER (CM)	DEFLECTION (SAG) OF CENTER OF LADDER SUPPORTING 1 PENNY (CM)
1.5	
2.5	
3.5	
4.5	
5.5	

Name

2. Graph your data from Table 2.1 on the grid below. Remember to label the axes and give the graph a title. Use the rubric provided by your teacher to assess your work. Then use your graph to answer the questions that follow.

3. Describe the relationship between ladder width and the amount the ladder sags.

_____

_____

4. If you had done the experiment with a 3-cm-wide ladder, how much do you think it would have sagged with a load of 1 penny on top of it? Explain.

_____

_____

5. Your final design must hold 10 pennies without sagging too much. Do you think that one of the ladder-bridges you tested would make a good final design? If yes, explain why. If no, what size width do you think would be able to support 10 pennies?

_____

_____

Teacher Page

2. RESEARCH THE PROBLEM: CREVASSE CRISIS!
RESEARCH PHASE 3: HOW DOES THE THICKNESS AFFECT BENDING?

OBJECTIVES

Students will:
- conduct a second controlled experiment
- collect experimental data in a table

MATERIALS

For each team:
- 2 identical textbooks
- a flat table to work on
- 3 pennies
- 1 piece of foam that is at least 30 cm $\times$ 17.5 cm
- 1 sheet of graph or white paper
- ruler (metric)
- scissors

BEFORE YOU TEACH

You may precut the "ladders" for your students or assign student helpers to cut ladders for each team before class. For Research Phase 3, each team needs five strips that are 30 cm long and 3.5 cm wide.

1. **CLASS** Introduce the next phase of research by discussing question 1. Explain that this activity is similar to the previous one except that instead of testing ladders of different widths, students will make various stacks of ladders to see how thickness affects the deflection (or sag) of the ladder-bridge. (Some students may want to spread out the three pennies.)

   ASK THE CLASS:

   Why is it important to keep the pennies in one stack and to place the stack in the middle of the ladder?

   Possible Answer(s): A stack of pennies more closely resembles a person's weight when the person is crossing the ladder. It is important to test the weakest part of the ladder, which is the middle of the ladder.

2. **TEAMS** Instruct teams to conduct the experiment and collect data. After that, students should graph the data and answer questions 3–5 on page 87. Give teams about 15 minutes to complete this activity.

   Note: If the ladder cannot support the weight of the three pennies and sags all the way to the bottom of the crevasse, there is no deflection to measure. Students can record "sagged to bottom."

Teacher Page *(continued)*

OBJECTIVES

Students will:

- produce and analyze a graph that relates two variables
- distinguish between independent and dependent variables
- connect the results of this research to the engineering design challenge
- compare rates of change (linear versus nonlinear relationships)

MATERIALS

For each class:

- the Rules of Thumb list

1. **TEAMS** As students are working on the graph and questions, use the questions below to help students.

   - As teams prepare to make the graph, ask each team: What is the best way to display this data? Should thickness be considered continuous? Could we have a thickness of 1.5 ladders?

     Possible Answer(s): It is difficult to cut a ladder in half along the thin side, but it is possible. It is fine for students to not connect the dots and leave it as a scatter plot.

   - To help teams answer question 4, ask them to compare the two graphs. Did one fall faster than the other? What does this mean about the effect of width versus thickness on deflection?

     Possible Answer(s): Thickness affects strength more than width. More specifically, width is related to strength linearly, while thickness is related to strength cubically. It is possible that the graphs might not show these relationships due to human error, measurement imprecision, and too few data points.

2. **CLASS** Provide the Rubric for Graphs on page 140 so students can assess their work. Go over questions 3–5 by having teams share their answers.

   - In question 4, some students may note that in the first experiment (testing width), they used only 1 penny, but in the second experiment (testing thickness), they used 3 pennies. Thus, students might intuitively reason based on their observations that increasing the thickness makes a greater difference because the sag is about the same (or perhaps less) as the sag in the first experiment, even though the weight is greater in the second experiment.

   - **ASK THE CLASS**
     Based on the data and your observations comparing the two experiments, should you use your ladders to increase the thickness or the width to create a strong bridge?

3. **CLASS** Review the design challenge, criteria, and constraints, and ask students what they learned from this research phase that they can add to the Rules of Thumb list.

> **ASSESSMENT**
> Use the Rubric for Graphs on page 140 to assess each team's graph.

2. RESEARCH THE PROBLEM: CREVASSE CRISIS!
RESEARCH PHASE 3: HOW DOES THE THICKNESS AFFECT BENDING?

1. Now that you have investigated ladder-bridge width, let's find out whether or not thickness might also be a factor. Do you think that the thickness of the ladder-bridge will affect how strong it is? Explain.

   _____

   _____

You will test how thickness affects strength by stacking ladders in the following experiment:

Step 1: Find your 3.5-cm-wide ladder from Research Phase 2. Cut out four more ladders that are the same size.

Step 2: Use the same book setup that was used in Research Phase 2, with a 12-cm gap between the books. Place one ladder across the "crevasse." Measure the thickness of the ladder. Hold a sheet of graph paper behind the ladder. Hold a book behind the graph paper for a sturdy marking surface. Make a mark to indicate the height of the bottom side of the ladder without any weight on it.

Step 3: Gently place 3 pennies, in a single stack, in the middle of the ladder. Hold the graph paper up again and mark the new height of the ladder, at its lowest point. Measure the distance between the two marks. This is the amount of deflection, or sag, of the ladder. Record the value in Table 2.2 below.

Step 4: Remove the pennies, and pile another ladder on top of the ladder that is already there, to make a thicker bridge. Repeat steps 2 and 3 for bridges that are 2, 3, 4, and 5 ladders thick.

Table 2.2: Thickness versus Deflection of Ladder-Bridge Supporting 3 Pennies

THICKNESS OF LADDER-BRIDGE (NUMBER OF LADDERS/CM)	DEFLECTION (SAG) OF LADDER-BRIDGE SUPPORTING 3 PENNIES (CM)
1 ladder/_____ cm	
2 ladders/_____cm	
3 ladders/_____cm	
4 ladders/_____cm	
5 ladders/_____cm	

© Museum of Science (Boston), Wong, Brizuela

2. Graph your data from Table 2.2 on the grid below. Remember to label the axes and give the graph a title. Use the rubric provided by your teacher to assess your work. Then use your graph to answer the questions that follow.

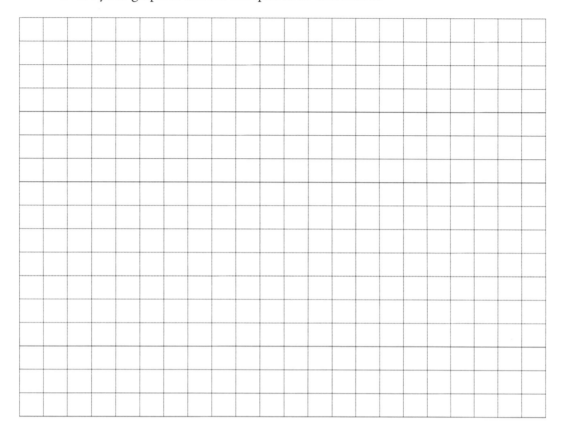

3. Describe the relationship between the thickness (number of ladders) of the bridge and how much it sags.

    _____

    _____

4. Based on these two experiments, which dimension of the ladder had a bigger effect on how much it sags: width or thickness? Explain.

    _____

    _____

5. Based on your results, can you explain why the craft stick broke more easily along its wide side than along its thin side?

    _____

    _____

Teacher Page

2. RESEARCH THE PROBLEM: CREVASSE CRISIS!
RESEARCH PHASE 4: HOW MIGHT OTHER SHAPES AFFECT BENDING?

OBJECTIVE: Students will explore the effect of shape on strength by constructing and testing beams.

MATERIALS

For each team:
- foam from previous two research phases
- tape
- scissors
- 2 identical textbooks
- pennies
- 1 sheet of graph or white paper
- ruler (metric)

For each class:
- the Rules of Thumb list

1. **CLASS** Explain that so far, students have investigated how width and thickness affect deflection of a bridge. Now they will investigate how shape affects deflection. Direct students to read the instructions on page 89. Hold a brief class discussion by asking the class the following questions:
   - Which of these beam shapes do you think is the strongest? Why do you think it is the strongest?
   - How does your research about the effect of width and thickness on strength relate to these shapes?

   Possible Answer(s): All the beams contain a vertical section that makes the beam thicker.

2. **TEAMS** Teams may use foam ladders from the previous two research phases to build beams of different shapes and compare their deflection. Inform teams that they don't have time to build all the different kinds of beams. Give teams about 10 minutes to experiment with beams of different shapes.

3. **CLASS** Wrap up this activity by asking teams the following questions:
   - Which beam shape(s) turned out to be strong? Which beam shapes were weak? What makes a beam strong or weak?
   - What is the advantage of using these shapes instead of simply stacking ladders on top of one another?

   Possible Answer(s): You use less material so the bridge would be lighter but still just as strong.

4. **CLASS** Review the design challenge criteria and constraints, and ask students what they learned from this research phase that they can add to the Rules of Thumb list.

2. RESEARCH THE PROBLEM: CREVASSE CRISIS!
RESEARCH PHASE 4: HOW MIGHT OTHER SHAPES AFFECT BENDING?

While acclimatizing to the altitude, one of your mountain guides finds some papers in a bag that might be helpful. When this guide isn't climbing Mount Everest, she works as an engineer and happens to have some drawings from work. They show some commonly shaped beams used in the construction industry. Some of these shapes may be helpful for making your ladder-bridge stronger. Use these pictures to think about how you might be able to cut or combine the ladders you are given to make the strongest possible ladder-bridge.

Which of these beams do you think would be the strongest? How do these shapes relate to your research comparing the effect of thickness versus width on strength? Write your ideas below. You may want to actually build some of these beams and test how many pennies they can hold.

_____

_____

_____

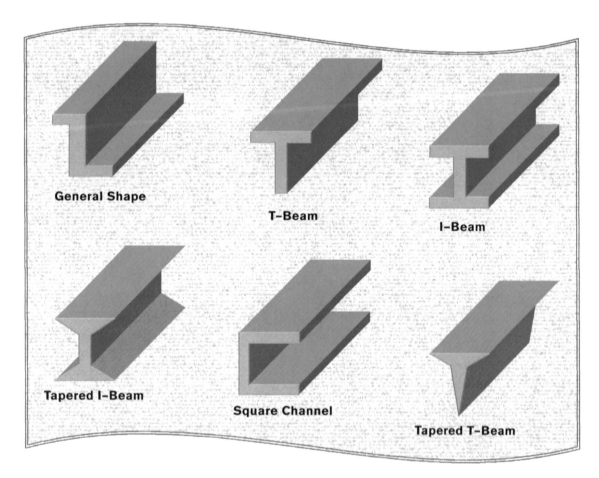

3. BRAINSTORM POSSIBLE SOLUTIONS: CREVASSE CRISIS!

OBJECTIVE: Students will individually brainstorm and sketch a bridge design.

MATERIALS

For each class:

- the Rules of Thumb list

1. [**CLASS**] Explain that students have completed the research phase and are moving on to the brainstorm phase. Together, read number 1 and review the engineering criteria.

2. [**INDIVIDUALS**] Instruct students to individually brainstorm for about 8 minutes. Students should review the research results and the Rules of Thumb list, sketch a design of a ladder-bridge that they think will meet the criteria, and label the diagram with dimensions.

3. BRAINSTORM POSSIBLE SOLUTIONS: CREVASSE CRISIS!

INDIVIDUAL DESIGN

1. You've done some great research and are ready to create some possible ladder-bridge designs. As you brainstorm possible solutions, keep the engineering criteria in mind. Look back at your answers to questions 1 and 2 on page 75, and fill in the blanks for the engineering criteria for your scale model in the box below.

ENGINEERING CRITERIA
STRONG ⟶ The ladder-bridge must be able to support 10 pennies without sagging more than _____ cm.
MATERIAL USE ⟶ Use as few ladders as possible, and no more than 5 ladders total.
WIDE TOP ⟶ The top of the ladder-bridge must be at least _____ cm wide to ensure that you can safely walk across.

2. Brainstorm on your own for a few minutes. Using the results of your team's research about bridge width, thickness, and shape, come up with one possible bridge design. Sketch your idea in the space below. Remember to label the number of ladders and dimensions (length, width, and thickness).

4. CHOOSE THE BEST SOLUTION: CREVASSE CRISIS!

OBJECTIVE: Students will share their individual brainstorm designs and decide on a team design.

MATERIALS

For each team:
- 5 pieces of foam that are each 30 cm × 2.4 cm
  OR
- 1 piece of foam that is at least 30 cm × 12 cm

For each class:
- the Rules of Thumb list

1. **TEAMS** Instruct students to take turns sharing their ideas and decide on a team design that they want to build and test. Students may have access to materials but are not to cut or tape the ladders yet. Remind students to review the Rules of Thumb list to help them make design decisions to meet criteria and constraints.

> **CLASSROOM MANAGEMENT TIPS**
> - If your students need more structure to share their individual ladder-bridge designs from the Brainstorm step, go over a list of things for each person to share (e.g., drawing of the bridge, the rationale for shape, the number of ladders, and the placement of ladders); a set time for each person to share (about 1 minute); and behavior expectations for teammates during sharing (e.g., no interruptions, no comments until everyone has shared, active listening).
> - Discuss with students how to come to an agreement on a team design. Team members can take turns discussing pros and cons of each design, identify commonalities in designs, compromise on areas of disagreement, and either vote or try to reach consensus.

2. **CLASS** If you plan on using students' team designs as an assessment, review the Rubric for Engineering Drawings on page 142 with students.

> **ASSESSMENT**
> Use the Rubric for Engineering Drawings on page 142 to assess students' work.

INTERESTING INFO
Foam is a substance that is formed by trapping gas bubbles in a liquid or solid. From the early twentieth century, various types of specifically manufactured solid foams came into use. The low density of these foams made them excellent as thermal insulators and flotation devices; their lightness and compressibility made them ideal as packing material. Some liquid foams are also used to extinguish fires.

4. CHOOSE THE BEST SOLUTION: CREVASSE CRISIS!

TEAM DESIGN

1. Have each member of your team share his or her ideas with the group. For each idea, think about the following:

   - What do you like the most about the design?
   - Do you think the design will meet all of the engineering criteria and constraints?

2. As a group, decide on one "best" solution. Draw a sketch of the ladder-bridge design in the space below. Label the number of ladders used and the dimensions of the ladder-bridge (length, width, thickness). Show how your design addresses all the criteria and constraints. Use the rubric provided by your teacher to check the quality and completeness of your drawing.

5. BUILD A PROTOTYPE/MODEL: CREVASSE CRISIS!

OBJECTIVE: Students will build their teams' bridge designs.

MATERIALS

For each team:
- 5 pieces of foam that are each 30 cm × 2.4 cm
  OR
- 1 piece of foam that is at least 30 cm × 12 cm
- ruler (metric)
- scissors
- tape

BEFORE YOU TEACH

You may precut the "ladders" for your students or assign student helpers to cut ladders for each team before class. Each team needs five strips that are each 30 cm long and 2.4 cm wide.

CLASS Go over the directions on the next page. Emphasize that students only have the materials they are given to construct the ladder-bridge. They may use fewer than five ladders, but they will not receive more ladders if they make a mistake. Give students about 20 minutes to build their ladder-bridge. Provide the Rubric for Prototype/Model on page 143 so students can assess their work.

ASSESSMENT

Use the Rubric for Prototype/Model on page 143 to assess completeness and craftsmanship of students' models.

---

5. BUILD A PROTOTYPE/MODEL: CREVASSE CRISIS!

You will now construct a model of your team's ladder-bridge design. You may use any part of the 5 foam ladders you were given and tape to construct your prototype. Write down any changes you make to your original design as you build your model. Please set aside any whole ladders that you do not use. Use the rubric provided by your teacher to assess your work.

6. TEST YOUR SOLUTION: CREVASSE CRISIS!

OBJECTIVE: Students will follow testing procedures to test their bridge prototypes.

MATERIALS

For each team:

- 2 identical textbooks
- a flat table to work on
- 1 sheet of graph or white paper
- ruler (metric)
- large cup with string attached
- 10–500 pennies (can be shared with other groups)

TEAMS During the testing phase, students will follow the instructions on the next page. To find the actual width of their bridge and the actual sag, students will need to divide the centimeter measurement of the model bridge by 6. After completing the testing phase, they may use the remaining time to discuss the questions on page 100. Provide the Rubric for Test, Communicate, and Redesign Steps on page 144 so students can assess their work.

ASSESSMENT

Use the Rubric for Test, Communicate, and Redesign Steps on page 144 to assess how well students followed testing procedures.

---

6. TEST YOUR SOLUTION: CREVASSE CRISIS!

You may use the rubric provided by your teacher to assess your work on the next few pages.

MATERIAL USED

How many of the 5 ladders did you need to use in order to build your design? If you used only part of one of the ladders, you must count this as a whole ladder. _____ ladders
Compare this to the number of ladders your classmates used in their designs.

WIDTH TEST

Measure the width of your ladder-bridge at its narrowest point: _____ cm
Using the scale you found in question 1 on page 75, find the actual width of your bridge: _____ m

Would your actual bridge be at least 0.4 meters wide?   ☐ yes ☐ no

STRENGTH TEST

Step 1: As you did during your research, set up two identical books on your desk, with 12 cm between them. Place your ladder-bridge model across the gap.

Step 2: Hold a sheet of graph paper (backed by a book for sturdiness) behind the ladder and mark the starting height of the bottom side of your ladder-bridge.

Step 3: Carefully place 10 pennies in a stack on top of the middle of your ladder-bridge.

Step 4: Place the same sheet of graph paper behind your sagging ladder-bridge. Make a mark to indicate the lowest point of the bottom side of your ladder-bridge.

Step 5: Measure the distance between the two marks—this is how much your ladder-bridge is sagging under the weight of the ten pennies.

Record the distance here: _____ cm

Step 6: Using the scale you found in question 1 on page 75, find the actual sag of your ladder-bridge: _____ m

Does your ladder-bridge hold 10 pennies without bending more than 0.2 meters? ☐ yes ☐ no

MAXIMUM STRENGTH TEST

Step 1: Set up two identical books on a desk so that a portion of each book is suspended over the edge of the desk. Place your ladder-bridge model across the gap over the edge of the desk.

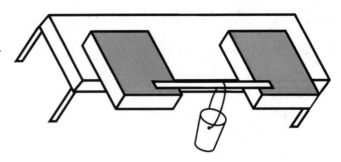

Step 2: Hang a large cup attached to a string around the middle of the ladder-bridge model.

Step 3: Gradually add pennies to the cup until the ladder-bridge model collapses.

Step 4: Count the number of pennies and record the number here: _____ pennies

7. COMMUNICATE YOUR SOLUTION: CREVASSE CRISIS!

OBJECTIVE: Students will answer questions to reflect on their designs and present their designs to the class.

1. **TEAMS** Teams should discuss and complete questions 1–4 as they prepare to present their design, test results, and reflections to the class.

2. **CLASS** To compare designs for the maximum strength test, complete a table like the one below. The higher the ratio, the higher the efficiency of the bridge (the more weight held per quantity of materials).

TEAM NAME	NUMBER OF PENNIES	NUMBER OF LADDERS	RATIO OF PENNIES TO LADDERS

3. **TEAMS** Teams should give presentations that are 2 to 3 minutes long. Encourage the other teams to ask questions and give suggestions for improving their design.

COMMUNICATE

7. COMMUNICATE YOUR SOLUTION: CREVASSE CRISIS!

1. Do you think that your ladder-bridge design was successful? Did it meet all of the criteria and constraints? Explain your answer fully.

    _____

    _____

    _____

2. Specifically, what are some strengths or advantages of your design? Explain.

    _____

    _____

    _____

3. What are some drawbacks or disadvantages of your design? Explain.

    _____

    _____

    _____

4. Would you feel safe crossing an actual crevasse using the ladder-bridge you designed? Explain.

    _____

    _____

    _____

Be prepared to present your answers to questions 1–4 to the class.

Teacher Page

8. REDESIGN AS NEEDED: CREVASSE CRISIS!

OBJECTIVE: Students will answer questions to consider how they can redesign their bridges.

1.   **TEAMS** Explain that designs can always be improved. Assign teams 7 to 8 minutes to answer questions 1 and 2.

2.   **CLASS** Wrap up the activity by asking the class the following questions:

   • What are the factors that affect the sag or deflection of a bridge?

   Possible Answer(s): The factors include width, thickness, and shape.

   • How did you investigate these factors?

   Possible Answer(s): We investigated width and thickness by building various bridge models of varying widths and thicknesses, and testing them to see how they sag. We investigated shape by building bridge models of different shapes to see how they sag.

   • How did you apply the results of your research to your ladder-bridge design?

   Possible Answer(s): We learned that thickness affects the sag more than width, so we focused on making the bridge thicker rather than wider.

   • What math skills did you use in this activity?

   Possible Answer(s): We used proportions to calculate the dimensions of our scale model. We made a line graph from our experimental data and analyzed the graph for trends and relationships. We drew a design that included labeled dimensions. We used rulers to measure the materials to build our design.

8. REDESIGN AS NEEDED: CREVASSE CRISIS!

1. Based on the test of your ladder-bridge design, what changes could you make to improve it?

    _____

    _____

    _____

    Explain how these changes would improve your design.

    _____

    _____

    _____

2. Consider the other teams' designs. How are their designs different from yours? Did the other teams use more or fewer ladders than your team? Did their ladder-bridges sag more or less than your bridge? What are the specific advantages and disadvantages of some of the other teams' designs?

    _____

    _____

    _____

    What have you learned from the other teams' designs that could help you improve your design?

    _____

    _____

    _____

INDIVIDUAL SELF-ASSESSMENT RUBRIC: CREVASSE CRISIS!

OBJECTIVE: Students will use a rubric to individually assess their involvement and work in this design challenge.

ASSESSMENT

Assign this reflection exercise as homework. You can write your comments on the lines below the self-assessment and/or use this in conjunction with the Student Participation Rubric on page 145.

TEAM EVALUATION: CREVASSE CRISIS!

OBJECTIVE: Students will evaluate and discuss how well they worked in teams.

ASSESSMENT

1. **INDIVIDUALS** Assign this reflection exercise as homework or during quiet classroom time.

2. **TEAMS** Instruct students to share their team evaluation reflections with one another and discuss how they can improve their teamwork during the next activity.

3. **CLASS** Point out any good examples of teamwork and areas to improve during the next activity.

STUDENT PAGE

INDIVIDUAL SELF-ASSESSMENT RUBRIC: CREVASSE CRISIS!

Use this rubric to reflect on how well you met behavior and work expectations during this activity. Check the box next to each expectation that you successfully met.

LEVEL 1	LEVEL 2	LEVEL 3	LEVEL 4	BONUS POINTS
Beginning to meet expectations	Meets some expectations	Meets expectations	Exceeds expectations	
☐ I was willing to work in a group setting.	☐ I met all of the Level 1 requirements.	☐ I met all of the Level 2 requirements.	☐ I met all of the Level 3 requirements.	☐ I helped resolve conflicts on my team.
☐ I was respectful and friendly to my teammates.	☐ I recorded the most essential comments from other group members.	☐ I made sure that my team was on track and doing the tasks for each activity.	☐ I helped my teammates understand the things that they did not understand.	☐ I responded well to criticism.
☐ I listened to my teammates and let them fully voice their opinions.	☐ I read all instructions.	☐ I listened to what my teammates had to say and asked for their opinions throughout the activity.	☐ I was always focused and on task: I didn't need to be reminded to do things; I knew what to do next.	☐ I encouraged everyone on my team to participate.
☐ I made sure we had the materials we needed and knew the tasks that needed to be done.	☐ I wrote down everything that was required for the activity.	☐ I actively gave feedback (by speaking and/or writing) to my team and other teams.	☐ I was able to explain to the class what we learned and did in the activity.	☐ I encouraged my team to persevere when my teammates faced difficulties and wanted to give up.
	☐ I listened to instructions in class and was able to stay on track.	☐ I completed all the assigned homework.		☐ I took advice and recommendations from the teacher about improving team performance and used feedback in team activities.
	☐ I asked questions when I didn't understand something.	☐ I was able to work on my own when the teacher couldn't help me right away.		☐ I worked with my team outside of the classroom to ensure that we could work well in the classroom.
		☐ I completed all the specified tasks for the activity.		

Approximate your level based on the number of checked boxes: _____ Bonus points: _____

Teacher comments: _____

TEAM EVALUATION: CREVASSE CRISIS!

How well did your team work together to complete the design challenge? Reflect on your teamwork experience by completing this evaluation and sharing your thoughts with your team. Celebrate your successes and discuss how you can improve your teamwork during the next design challenge.

Rate your teamwork. On a scale of 0–3, how well did your team do? 3 is excellent, 0 is very poor. Explain how you came up with that rating. Was it the same, better, or worse than the last activity?

List things that worked well. Example: We got to our tasks right away and stayed on track.

List things that did not work well. Example: We argued a lot and did not come to a decision that everyone could agree on.

How can you improve teamwork? Make the action steps concrete. Example: We need to learn how to make decisions better. Therefore, I will listen and respond without raising my voice.

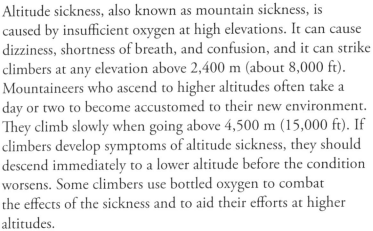

Design Challenge 3
Sliding Down!
INTRODUCTION

OBJECTIVE: Students will read and understand the problem presented for the third design challenge.

 **CLASS** Together, read the introduction on the next page.

ASK THE CLASS:

- What ideas do you have for getting the sick climbers safely back to Camp IV? (You can ask them to first brainstorm in teams and then share with the whole class.)

INTERESTING INFO

Altitude sickness, also known as mountain sickness, is caused by insufficient oxygen at high elevations. It can cause dizziness, shortness of breath, and confusion, and it can strike climbers at any elevation above 2,400 m (about 8,000 ft). Mountaineers who ascend to higher altitudes often take a day or two to become accustomed to their new environment. They climb slowly when going above 4,500 m (15,000 ft). If climbers develop symptoms of altitude sickness, they should descend immediately to a lower altitude before the condition worsens. Some climbers use bottled oxygen to combat the effects of the sickness and to aid their efforts at higher altitudes.

Design Challenge 3

Sliding Down!
INTRODUCTION

You and your climbing team have made it to the summit of Mount Everest! You're standing on the top of the world! Unfortunately, at this high altitude, the air is only one third as dense as it is at sea level. This means that with every breath, you are taking in much less oxygen than you need in order to function. Some of your teammates are showing signs of altitude sickness. They are dizzy, tired, and very confused. A few are even beginning to hallucinate. The only way to help them feel better is to get them down to a lower altitude. Unfortunately, in their oxygen-deprived state, they will have a hard time traveling the steep, rocky terrain back down to Camp IV.

Luckily, you have not been as affected by the high altitudes as some of the others. What can you do to help them? Is there a way to get all of the sick climbers safely back to Camp IV?

Teacher Page

1. DEFINE THE PROBLEM: SLIDING DOWN!

OBJECTIVE: Students will read and understand the criteria and constraints of the design challenge.

CLASS Together, read the first paragraph on the next page, which describes one possible solution—a kind of zip-line transportation system. This activity challenges students to design a model zip-line that can transport small figurines by allowing gravity to lower the transport down along a fishing line. Make sure that students understand the three criteria and show them the materials that they will be using to build the zip-line system.

ASK THE CLASS:

- How can you control how fast the transport carrier moves along the zip-line?
  Possible Answer(s): You can vary the angle of the line to the mountain. The transporter moves faster when the line is steeper.

Explain that students will start the research phase by investigating how the angle of the zip-line to the mountain affects the speed of the transporter.

INTERESTING INFO

PROBLEM SOLVING:
You have a teammate who is affected by altitude sickness, and his body is slowly losing oxygen. Luckily, there's an equipment store even at this altitude! You want to go to the store and rent an oxygen tank for your teammate, so he can recover while you wait for others to come to the rescue. The rescue team is on the way, but it will be another hour before they arrive. You have $150, and you can only carry back two tanks with you. Choose the two oxygen tanks that will last your teammate until help comes. What are all the possible solutions? Which solution is the best? Explain.

OXYGEN TANK	PRICE	DURATION
Tank 1	$ 45	10 minutes
Tank 2	$ 50	20 minutes
Tank 3	$ 65	30 minutes
Tank 4	$ 70	40 minutes
Tank 5	$ 95	50 minutes

1. DEFINE THE PROBLEM: SLIDING DOWN!

Since you are heading back down the mountain, gravity is on your side. There may be a way to take advantage of this fact and transport your sick teammates back to Camp IV. Although he has never done this before, or even seen it done, one of the Sherpa guides suggests that you design some kind of zip-line transportation system. With an effective design, the sick climbers could zip down the mountainside without having to exert much thought or effort. Your task is to design this transportation system. It should travel fast enough so that your sick teammates can start feeling better more quickly, but it should be slow enough so that it is safe for passengers. The transporter may hang from the zip-line, or it may sit on a platform that runs on top of the zip-line.

ENGINEERING CRITERIA

ADEQUATE SPEED ⟶ The transporter must move along the zip-line at a speed that is between 1 meter per second and 1.5 meters per second.

SAFE ⟶ The transporter must be able to safely carry two people (you will test with toy figures) along the full length of the zip-line without them falling off.

RETURN ⟶ The design must include some way for the transporter to be sent quickly back up to the summit, since only two people at a time can ride.

ENGINEERING CONSTRAINTS

- Your zip-line transportation system must span the full distance from the summit (8,850 m above sea level) to Camp IV (7,925 m above sea level).
- You will have the following materials available to you:

ACTUAL MATERIALS	MATERIALS THAT WILL REPRESENT THE ACTUAL MATERIALS
heavy-duty cables for the zip-line	fishing line
wooden planks to make the transporter	cardboard
hammer and nails	scissors and tape
extra ladder parts to support the cables	metersticks

Teacher Page

2. RESEARCH THE PROBLEM: SLIDING DOWN!
RESEARCH PHASE 1: INVESTIGATING ANGLES TO FIND AN ADEQUATE SPEED

OBJECTIVES

Students will:
- conduct a controlled experiment
- measure angles using a protractor

MATERIALS

For each team:
- 2 metersticks
- piece of fishing line 2.05 m long
- piece of straw 5 cm long
- scissors
- protractor
- ruler (metric)
- stopwatch (measures at least to nearest tenth of a second)
- half sheet of chart paper
- markers

1. **CLASS** Together, read the first paragraph on page 114, which asks students to think about how angle affects speed. Answer the questions as a group.

    ASK THE CLASS:
    - Is there an angle at which the transporter wouldn't even move?

      Possible Answer(s): Yes, as the angle approaches 0° (parallel to the ground), the transporter would move slower and slower until the transporter could not overcome the friction of the line to slide down.

2. **TEAMS** Demonstrate the experiment setup and go over the directions. Instruct students to position the metersticks along the wall so that the 0 cm mark is on the floor. The fishing line must be held tight and kept motionless during each trial. Explain that the fishing line is 2.05 m because the 0.05 m is for the 5-cm-long straw to travel exactly 2 m. Emphasize that it is extremely important for the Timer to start the stopwatch at the moment the straw starts moving and stop the stopwatch at the moment the straw reaches the end of the line. Give teams about 10 minutes to complete this experiment. Give each team half a sheet of chart paper on which to write their data results.

© Museum of Science (Boston), Wong, Brizuela

Teacher Page *(continued)*

OBJECTIVES

Students will:
- compare and discuss appropriate measures of central tendency (mean, median, mode)
- apply the distance-time-speed formula

MATERIALS

For each team:
- calculators

1.   **CLASS**   Post all the results in one location. Discuss question 1 as a class.

ASK THE CLASS:

- Is there a better way to measure travel time? How can we decrease human error?

  Possible Answer(s): The End Holder could yell out "stop" as soon as he feels the straw. Feeling the straw as it arrives at the bottom of the string might be more accurate than seeing it. A more high-tech way might involve using motion detectors that are connected to a timer. The detectors could trigger the timer to start when the straw starts moving and stop the timer when the straw reaches the bottom of the string and stops.

- Before students calculate the class averages in number 2, you might want to review different measures of central tendency (mean, median, and mode). This is especially important if the teams' data include outliers (data that seem to be very different from the others).

ASK THE CLASS:

- Which of the measures of central tendency best represent the "typical" value given a particular group of data?

  Possible Answer(s): For instance, let's say that four teams got the following values for angle measurements when the height of the zip-line is 0.4 m: 10°, 15°, 19°, 32°. A line plot showing the range and spread of data also shows that 32° is an outlier:

  The mean is 19, the median is 17, and there is no single mode. In this case, the median is probably a better "average" than the mean because the outlier pulls up the mean—allowing one data point an inordinate amount of influence.

2.   **TEAMS**   Instruct teams to complete numbers 2 and 3.

OBJECTIVES

Students will:

- produce and analyze a graph that relates two variables
- locate and represent the range of acceptable values on a graph to meet a design criteria
- distinguish between independent and dependent variables
- connect the results of this research phase to solving the design challenge

MATERIALS

For each class:

- a sheet of chart paper for the Rules of Thumb list (see page 15)

1. **CLASS** You can either choose to guide students through setting up this graph or assign this task as an assessment.

   ASK THE CLASS:

   - For this graph showing angle versus speed, which variable is independent and which is dependent? Why?

     Possible Answer(s): The angle is the independent variable because that is what we controlled. The speed is the dependent variable because it is dependent on the angle.

   - Should you use a line graph to represent the data? Why or why not?

     Possible Answer(s): We should use a line graph because the angle represents continuous data—there are an infinite number of values in between each interval of angle measurements we used.

   - What do you expect the graph to look like?

     Possible Answer(s): The line should rise as it moves to the right. The speed increases as the angle increases.

HELPFUL EXTENSION

Students might find it helpful to make another graph to show height versus speed because height is the variable that is adjusted in the experiment that affected the time (speed). The height versus speed graph might make it easier for students to make design decisions to successfully meet the speed criteria.

---

2. **TEAMS** Instruct teams to make their graph and answer questions 5–8.

3. **CLASS** Debrief the graphing exercise by discussing the following:
   - How can we represent the range of acceptable angle values on the graph?

     Possible Answer(s): Draw two horizontal lines on the graph to represent the minimum and maximum speeds specified in the design criteria to help students visualize the range of possible angles.

   - What if we graphed angle versus time? As angle increases, what would happen to the time it takes for the straw to slide down the line? What would the graph look like?

     Possible Answer(s): The graph would start high but decrease as it moves to the right. As the angle increases, the time decreases, which means that speed is increasing, given a constant distance.

   - What might the graph look like if you collected and plotted data for all angles from $0°$ to $90°$? Particularly, what would speed look like for very large angles?

     Possible Answer(s): The graph would start at 0 for speed, because at the 0° angle, the line would be parallel to the floor and the straw would not move. As the angle increases, the speed would pick up. As it nears 90°, the speed would be nearly equivalent to free fall, which is determined by gravity and the friction of the air.

4. **CLASS** Review the design challenge, criteria, and constraints and ask students what they learned from this research phase that they can add to the Rules of Thumb list.

ASSESSMENT
Use the Rubric for Graphs on page 140 to assess each team's graph.

OPTIONAL CCSS ENHANCEMENTS
To address additional aspects of the Common Core State Standards, direct students to use variables to write an equation when calculating speed. Have students calculate and discuss the outliers from class data. Show students a best-fit line of the average speed data and discuss this relationship.

Name
STUDENT PAGE

2. RESEARCH THE PROBLEM: SLIDING DOWN!
RESEARCH PHASE 1: INVESTIGATING ANGLES TO FIND AN ADEQUATE SPEED

You will be investigating how angle affects the speed at which something moves along a zip-line. Before you begin your experiment, think about these questions: How do you think angle will affect speed? What do you think would happen if the zip-line were very steep? What if it were not steep at all?

Student jobs: Assign each member of your team one of the jobs listed below.

Start Holder: _____   **End Holder:** _____

Timer: _____   **Recorder:** _____

Step 1: **Start Holder**—Tape two metersticks end-to-end vertically against the wall, with the zero end on the floor or table. Thread the fishing line through the straw. Position the top end of the 2.05 m long fishing line at a height of 0.4 m. Then, hold the piece of straw at the top end of the fishing line.

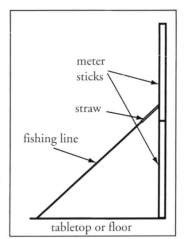

meter sticks

straw

fishing line

tabletop or floor

Step 2: **End Holder**—Pull the other end of the fishing line tight against the floor or table (where the zero mark of the bottom meterstick is resting), and hold it there. Use the entire length of the fishing line so that the distance that the straw travels is always the same.

Step 3: **Recorder**—Using a protractor, measure the angle between the end of the fishing line and the floor. Record this value in Table 3.1.

Step 4: **Timer**—Check the angle measurement. Have your stopwatch ready. Say, "One, two, three, release!" At the moment you say "release!" start your stopwatch.

Step 5: **Start Holder**—At the moment the **Timer** says "release!" let go of the straw.

Step 6: **Timer**—At the moment the straw reaches the end of the fishing line, stop your stopwatch, and read the time value aloud.

Step 7: **Recorder**—Record the time value in Table 3.1.

Step 8: Repeat steps 1–7 for the other heights listed in Table 3.1.

Table 3.1: Time for a Straw to Travel a 2-m Fishing Line at Different Angles

HEIGHT OF BEGINNING OF ZIP-LINE (M)	0.4	0.6	0.8	1.0	1.2	1.4
ANGLE (DEGREES)						
TIME (SECONDS)						

© Museum of Science (Boston), Wong, Brizuela

1. Each team will share their angle and time results with the rest of the class. Once you have seen all of the other groups' data, compare your results. Are the times about the same or very different? If they are different, why might this be the case?

    _____

    _____

2. Calculate the class averages of angle measurements and time data, and record these values in Table 3.2 at the bottom of the page.

3. Now that you have average time values, you can calculate the average speed at which the straw traveled down the 2-meter-long fishing wire. Using the formula for speed (see below), calculate the speed for each angle. Record your calculations in Table 3.2. Show your work.

    Speed = Distance/Time (Since the straw travels down the fishing wire 2 meters, distance = 2 m.)

Table 3.2: Class Average Data for Time to Travel Along Zip-Line at Different Angles

HEIGHT OF BEGINNING OF ZIP-LINE (M)	0.4	0.6	0.8	1.0	1.2	1.4
CLASS AVERAGE ANGLE (DEGREES)						
CLASS AVERAGE TIME (SECONDS)						
AVERAGE SPEED (METERS/SECOND)						

4. Plot a graph showing angle versus speed based on your data in Table 3.2. Remember to label the axes, and give the graph a title. Use the rubric provided by your teacher to assess your work. Then use your graph to answer the questions that follow.

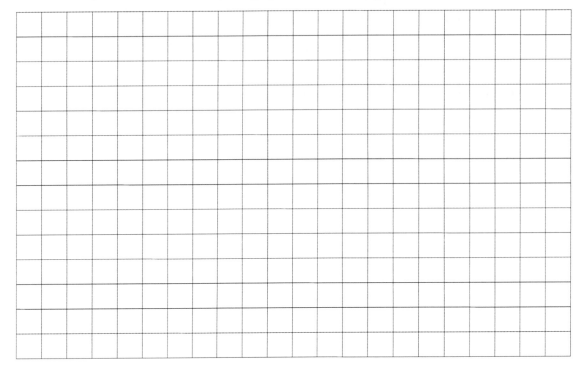

5. Describe the relationship between the angle of the fishing line and the speed of the straw.

_____

_____

6. How fast do you think the straw would travel if the fishing line were at an angle of 52°? Explain.

_____

_____

7. What range of angles will give an acceptable speed (between 1 m/s and 1.5 m/s)?

_____

8. What angle do you think would work best for your zip-line transportation system? Explain.

_____

_____

Teacher Page

2. RESEARCH THE PROBLEM: SLIDING DOWN!
RESEARCH PHASE 2: HOW CAN YOU MAKE THE TRANSPORTER SAFE?

OBJECTIVE: Students will brainstorm factors that affect the stability of the transporter.

MATERIALS

For each class:

- the Rules of Thumb list

1. **CLASS** Discuss students' ideas together.

2. **CLASS** Review the design challenge, criteria, and constraints and ask students what they learned from this research phase that they can add to the Rules of Thumb list.

INTERESTING INFO

Engineers use a lot of math in their work. The basic mathematical concepts and algebraic equations allow engineers to formulate theories and perform calculations. Using statistics and probability, engineers can test their hypotheses by analyzing data where samples tested may contain variability or where experimental error is likely to occur. Engineers use data analysis, such as filtering and coding information, to describe, summarize, and compare the data with their initial hypotheses. Finally, engineers use modeling and simulations to predict the behavior and performance of their designs before they are actually built.

2. RESEARCH THE PROBLEM: SLIDING DOWN!
RESEARCH PHASE 2: HOW CAN YOU MAKE THE TRANSPORTER SAFE?

In order for your zip-line transportation system to be considered safe, its passengers must be able to stay on the transporter with little effort. The sick people are very tired and disoriented, and may not be able to hold onto the transporter. Therefore, the transporter must be very stable so that no one falls off.

What factors do you think might affect the stability of the zip-line transporter? List at least four ideas.

Teacher Page

3. BRAINSTORM POSSIBLE SOLUTIONS: SLIDING DOWN!

OBJECTIVE: Students will review the design challenge's criteria and constraints and individually sketch a design of the zip-line transporter system.

MATERIALS

For each class:

- the Rules of Thumb list

1. **CLASS** Review the design criteria and have students think about other modes of transportation.

   ASK THE CLASS:

   - What makes other modes of transportation safe?

     Possible Answer(s): seat belts, fully enclosed cars, and so forth

   - How can you attach the string to the cart to pull it up the mountain? Where would you attach the other end of the string?

     Possible Answer(s): One way is to attach one end of the string to the back of the cart and the other end to the top of the mountain. When it's time to return the cart to the top, the people at the top of the mountain could simply drag the cart back up the mountain. In real life, it would be difficult to simply drag it up by a tether, so a pulley system might be implemented to alleviate the amount of force needed.

   Also explain that the return mechanism should be operable from either the top or the bottom. Students may not simply push the cart up to the top by hand. It might be helpful to discuss pulley systems briefly.

2. **INDIVIDUALS** Instruct students to individually brainstorm some ideas for the zip-line design and sketch their designs on page 120. Remind students to review the Rules of Thumb list when making design decisions.

3. BRAINSTORM POSSIBLE SOLUTIONS: SLIDING DOWN!

INDIVIDUAL DESIGN

You've done some great research and are ready to create some possible zip-line transportation system designs. Remember to keep the engineering criteria in mind.

ENGINEERING CRITERIA	
ADEQUATE SPEED ⟶	The transporter must move along the zip-line at a speed that is between 1 meter per second and 1.5 meters per second.
SAFE ⟶	The transporter must be able to safely carry two people (you will test with toy figures) along the full length of the zip-line without them falling off.
RETURN ⟶	The design must include some way for the transporter to be sent quickly back up to the summit, since only two people at a time can ride.

Brainstorm on your own for a few minutes, and think about how you will meet all of the criteria. Then, sketch your idea in the space below and label all parts of your design.

© Museum of Science (Boston), Wong, Brizuela

4. CHOOSE THE BEST SOLUTION: SLIDING DOWN!

OBJECTIVE: Students will share their individual brainstorm solutions and decide on a team zip-line transporter system.

MATERIALS

For each team:
- 2 metersticks
- piece of fishing line 5 m long
- piece of straw 5 cm long
- cardboard (letter-sized)
- small toy figures
- protractor

For each class:
- the Rules of Thumb list

TEAMS Instruct teams to share their ideas, choose the best parts, and combine them to create a team design that should be labeled with materials and dimensions. Give teams access to the materials so that they may see and feel them. However, students are not to cut or construct anything until the design is drawn and labeled. Remind students to review the Rules of Thumb list when making design decisions.

CLASSROOM MANAGEMENT TIPS
- If students need more structure to share their individual zip-line trasnportation designs from the Brainstorm step, go over a list of things for each person to share (e.g., the angle or the height of zip-line, the justification for choosing a particular angle or height, the transporter design, or the rationale for the transporter design); a set time for each person to share (about 1 minute); and behavior expectations for teammates during sharing (e.g., no interruptions, no comments until everyone has shared, active listening).

- Discuss with students how to come to an agreement on a team design. Team members can take turns discussing pros and cons of each design, identify commonalities in designs, compromise on areas of disagreement, and either vote or try to reach consensus.

ASSESSMENT

Introduce students to the Rubric for Engineering Drawings on page 142 by showing students examples of student work on pages 151–154 and using the rubric to grade each drawing. You can first assess one drawing with the whole class using the rubric and then have students work in pairs to assess the other drawings. Debrief as a whole class. Students can use the rubric to self-assess their own drawings of the zip-line design.

INTERESTING INFO

A zip-line consists of a pulley suspended on a cable mounted on an incline. A zip-line is designed to enable a user to move from the top to the bottom of the inclined rope or cable. The user does this by holding onto the freely moving pulley. Zip-lines come in all forms and shapes. A person can ride a small zip-line by holding onto the pulley and hanging underneath. Larger rides necessitate that the rider wear a safety harness. These rides can be very high, starting at a height of nearly 10 meters, and travel more than 60 meters. Users of zip-lines must have means of stopping themselves. Typical mechanisms include a mat or netting at the lower end of the incline. For safety purposes, proper knowledge of rope work is required in order to construct a zip-line.

4. CHOOSE THE BEST SOLUTION: SLIDING DOWN!

1. Have each member of your team share his or her ideas with the group. For each idea, think about the following:

   - What do you like the most about the design?
   - Do you think the design will meet all of the engineering criteria and constraints?

2. As a group, decide on one "best" solution. Draw a sketch of the zip-line transportation system in the space below. Then label all parts of the design, including the zip-line angle and dimensions. Show how your design addresses all the criteria and constraints. Use the rubric provided by your teacher to check the quality and completeness of your drawing. Practice using the rubric on sample drawings provided by your teacher. Use a protractor and construct accurate triangles.

Teacher Page

5. BUILD A PROTOTYPE/MODEL: SLIDING DOWN!

OBJECTIVE: Students will build a model of their zip-line transporter system.

MATERIALS
For each team:
- 2 metersticks
- piece of fishing line 5 m long
- piece of straw 5 cm long
- ruler (metric)
- protractor
- cardboard (letter-sized)
- scissors
- tape
- small toy figures

1.   **CLASS**   Tell students that they will build a model of their design.

2.   **TEAMS**   As you approve each team's design, give them the materials to build their zip-line models. Remind teams to write down any changes they made to their plan as they build. Provide the Rubric for Prototype/Model on page 143 so students can assess their work.

ASSESSMENT
Use the Rubric for Model/Prototype on page 143 to assess completeness and craftsmanship of students' models.

---

5. BUILD A PROTOTYPE/MODEL: SLIDING DOWN!

You will now construct a model of your team's zip-line transportation system. You may use fishing line, straws, cardboard, scissors, and tape to construct your prototype. Write down any changes to your original design as you build your model. Use the rubric provided by your teacher to assess your work.

6. TEST YOUR SOLUTION: SLIDING DOWN!

OBJECTIVE: Students will test their teams' zip-line transporter systems.

MATERIALS

For each team:
- stopwatch (or digital timer that displays fractions of seconds)
- calculator

TEAMS Instruct teams to conduct the three tests by following the directions on the next page. Provide the Rubric for Test, Communicate, and Redesign Steps on page 144 so students can assess their work.

ASSESSMENT

Use the Rubric for Test, Communicate, and Redesign steps on page 144 to assess how well students followed testing procedures.

HELPFUL EXTENSION

If timers that display fractions of seconds are available, have students practice division of fractions.

---

6. TEST YOUR SOLUTION: SLIDING DOWN!

You may use the rubric provided by your teacher to assess your work on the next few pages.

SPEED TEST

Student jobs: Assign each member of your team one of the jobs listed below.

Begin Holder: _____   **End Holder:** _____

Timer: _____   **Recorder:** _____

Step 1: Begin Holder—Your zip-line transportation system should already be set up. Hold the actual transporter at the top end of the fishing line.

Step 2: End Holder—Pull the other end of the fishing line tight against the floor or table, and hold it there.

Step 3: Timer—Have your stopwatch ready. Say, "One, two, three, release!" At the moment you say "release!" start your stopwatch.

Step 4: Begin Holder—At the moment the **Timer** says "release!" let go of the transporter.

Step 5: Timer—At the moment the straw reaches the end of the fishing line, stop your stopwatch and read the time value aloud.

Step 6: Recorder—Record the time here: _____ seconds

Step 7: Using the formula speed = distance/time, calculate the average speed of the transporter. How fast did your transporter travel along the fishing line?_____m/s?

Is the speed between 1 m/s and 1.5 m/s?   ☐ yes ☐ no

SAFETY TEST

Hold the transporter at the top, or beginning, of the fishing line. Place 2 toy figures on or inside the transporter. Let go of the transporter, and watch the 2 toy figures ride down the zip-line.

Did they remain on or inside the transporter for the entire ride?   ☐ yes ☐ no

RETURN MECHANISM TEST

Call your teacher over. Then, position the transporter at the bottom end of the fishing line. Demonstrate to your teacher how you can get the transporter back up to the top end of the fishing line without touching the transporter at any point. Now, ask your teacher to verify that you and your teammates were able to return the transporter to the "summit" without touching the transporter.   ☐ yes ☐ no

7. COMMUNICATE YOUR SOLUTION: SLIDING DOWN!

OBJECTIVE: Students will answer questions to reflect on their designs, present their designs to other teams, and comment on one another's designs.

MATERIALS (OPTIONAL)

For each team:
- chart paper or poster board
- pack of markers

1. **TEAMS** Teams should discuss and answer questions 1–4 as they prepare their presentations. You may want each team to prepare a poster to show a drawing of their design and the zip-line's test performance.

2. **CLASS** Invite teams to each spend 3 to 5 minutes presenting their design and results.

7. COMMUNICATE YOUR SOLUTION: SLIDING DOWN!

1. Do you think that your zip-line transportation system was successful? Did it meet all of the criteria and constraints? Explain.

_____

_____

_____

2. Specifically, what are some strengths or advantages of your design? Explain.

_____

_____

_____

3. What are some drawbacks or disadvantages of your design? Explain.

_____

_____

_____

4. Do you think that a zip-line transportation system is the best way to solve the problem of getting the sick climbers back to Camp IV? Can you think of a better way to get them down the mountain? Explain.

_____

_____

_____

Be prepared to present your answers to questions 1–4 to the class.

8. REDESIGN AS NEEDED: SLIDING DOWN!

OBJECTIVE: Students will answer questions to consider how they can redesign their zip-line transportation systems.

1. **TEAMS** Instruct teams to use what they learned from other teams' presentations of their designs—both successes and failures—to improve their own designs by answering questions 1 and 2.

2. **CLASS** Wrap up the activity by asking the class the following question:
   - How did you use math to solve the zip-line design problem?

     Possible Answer(s): We experimented with angles to see how fast the straw travels down the line so we could find the optimal angle. We also used the ruler to measure and construct the transporter for the zip-line.

REDESIGN

8. REDESIGN AS NEEDED: SLIDING DOWN!

1. Based on the test of your zip-line transportation system, what changes could you make to improve your design?

_____

_____

_____

Explain how these changes would improve your design.

_____

_____

_____

2. Identify one thing that you learned from another group's design that you can use to improve your design.

_____

_____

_____

Explain how this will improve your design.

_____

_____

_____

INDIVIDUAL SELF-ASSESSMENT RUBRIC: SLIDING DOWN!

OBJECTIVE: Students will use a rubric to individually assess their involvement and work in this design challenge.

> **ASSESSMENT**
>
> Assign this reflection exercise as homework. You can write your comments on the lines below the self-assessment and/or use this in conjunction with the Student Participation Rubric on page 145.

TEAM EVALUATION: SLIDING DOWN!

OBJECTIVE: Students will evaluate and discuss how well they worked in teams.

> **ASSESSMENT**
>
> 1. **INDIVIDUALS** Assign this reflection exercise as homework or during quiet classroom time.
>
> 2. **TEAMS** Instruct students to share their team evaluation reflections with one another.
>
> 3. **CLASS** Point out any good examples of teamwork.

INDIVIDUAL SELF-ASSESSMENT RUBRIC: SLIDING DOWN!

Use this rubric to reflect on how well you met behavior and work expectations during this activity. Check the box next to each expectation that you successfully met.

LEVEL 1 Beginning to meet expectations	LEVEL 2 Meets some expectations	LEVEL 3 Meets expectations	LEVEL 4 Exceeds expectations	BONUS POINTS
☐ I was willing to work in a group setting.	☐ I met all of the Level 1 requirements.	☐ I met all of the Level 2 requirements.	☐ I met all of the Level 3 requirements.	☐ I helped resolve conflicts on my team.
☐ I was respectful and friendly to my teammates.	☐ I recorded the most essential comments from other group members.	☐ I made sure that my team was on track and doing the tasks for each activity.	☐ I helped my teammates understand the things that they did not understand.	☐ I responded well to criticism.
☐ I listened to my teammates and let them fully voice their opinions.	☐ I read all instructions.	☐ I listened to what my teammates had to say and asked for their opinions throughout the activity.	☐ I was always focused and on task: I didn't need to be reminded to do things; I knew what to do next.	☐ I encouraged everyone on my team to participate.
☐ I made sure we had the materials we needed and knew the tasks that needed to be done.	☐ I wrote down everything that was required for the activity.	☐ I actively gave feedback (by speaking and/or writing) to my team and other teams.	☐ I was able to explain to the class what we learned and did in the activity.	☐ I encouraged my team to persevere when my teammates faced difficulties and wanted to give up.
	☐ I listened to instructions in class and was able to stay on track.	☐ I completed all the assigned homework.		☐ I took advice and recommendations from the teacher about improving team performance and used feedback in team activities.
	☐ I asked questions when I didn't understand something.	☐ I was able to work on my own when the teacher couldn't help me right away.		☐ I worked with my team outside of the classroom to ensure that we could work well in the classroom.
		☐ I completed all the specified tasks for the activity.		

Approximate your level based on the number of checked boxes: _____   Bonus points: _____

Teacher comments: _____

TEAM EVALUATION: SLIDING DOWN!

How well did your team work together to complete the design challenge? Reflect on your teamwork experience by completing this evaluation and sharing your thoughts with your team. Celebrate your successes!

Rate your teamwork. On a scale of 0–3, how well did your team do? 3 is excellent, 0 is very poor. Explain how you came up with that rating. Was it the same, better, or worse than the last activity?

List things that worked well. Example: We got to our tasks right away and stayed on track.

List things that did not work well. Example: We argued a lot and did not come to a decision that everyone could agree on.

How can you improve teamwork? Make the action steps concrete. Example: We need to learn how to make decisions better. Therefore, I will listen and respond without raising my voice.

Resources & Appendices

- EDP: Engineering Design Process
- Math & Engineering Concepts
- Important Vocabulary Terms
- Rubrics
- Student Work Samples

EDP: ENGINEERING DESIGN PROCESS

Engineers all over the world have one thing in common. They use the engineering design process (EDP) to solve problems. These problems can be as complicated as building a state-of-the-art computer or as simple as making a warm jacket. In both cases, engineers use the EDP to help solve the problems. Although engineers may not strictly follow every step of the EDP in the same order all the time, the EDP serves as a tool that helps to guide engineers in their thinking process and approach to a problem. Below is a brief outline of each step.

DEFINE	The first step is to **define** the problem. In doing so, remember to ask questions! What is the problem? What do I want to do? What specifications should my solution meet to successfully solve the problem (also called "criteria")? What factors may limit possible solutions to this problem (also called "constraints")?
RESEARCH	The next step is to conduct **research** on what can be done to solve the problem. What are the possible solutions? What have others already done? Use the Internet and the library to conduct investigations and talk to experts to explore possible solutions.
BRAINSTORM	**Brainstorm** ideas and be creative! Think about possible solutions in both two and three dimensions. Let your imagination run wild. Talk with your teacher and fellow classmates.

4 CHOOSE	**Choose** the best solution that meets all the criteria and constraints. Any diagrams or sketches will be helpful for later engineering design steps. Make a list of all the materials the project will need.
5 BUILD	Use your diagrams and list of materials as a guide to **build** a model or prototype of your solution.
6 TEST	**Test** and evaluate your prototype. How well does it work? Does it satisfy the engineering criteria and constraints?
7 COMMUNICATE	**Communicate** with your peers about your prototype. Why did you choose this design? Does it work as intended? If not, what could be fixed? What were the trade-offs in your design?
8 REDESIGN	Based on information gathered in the testing and communication steps, **redesign** your prototype. Keep in mind what you learned from one another in the communication step. Improvements can always be made!

MATH AND ENGINEERING CONCEPTS

In these three *Everest Trek* activities, you will integrate engineering with math to solve problems and design prototypes. Specifically, you will:

- collect, represent, and analyze data
- use various measuring tools to find length, area, and angle measurements
- solve problems involving proportions and scaling, and build scale models
- investigate the relationships between sets of variables, and specifically, how a change in one variable affects another variable
- apply the engineering design process to solve problems

IMPORTANT VOCABULARY TERMS

ACCLIMATIZE
to physically adapt to changes in your environment, such as changes in temperature and altitude

ALTITUDE SICKNESS
dizziness, nausea, loss of concentration and balance, and possible hallucinations caused by a lack of oxygen at high altitudes

ASCENT
an upward slope, or incline

BASE CAMP
a climbing expedition's main headquarters and starting point, usually near the bottom of the mountain

CREVASSE
a deep split in a large mass of ice or glacier

DEFLECTION
the bending of a structure away from its resting point, due to stress

DESCENT
a downward slope

ENGINEERING
the applications of math and science to practical ends, such as design or manufacture

ENGINEERING CONSTRAINTS
limiting factors to consider when designing a model

ENGINEERING CRITERIA

specifications met by a successful solution

ICEFALL

a part of a mass of ice or glacier that has broken off

INSULATOR

a material that prevents heat, electricity, or sound from passing through it

SHERPAS

people native to the Himalayan region (Because they have always lived in high altitudes, Sherpas are excellent mountaineers and often lead mountain-climbing expeditions.)

SUMMIT

the highest point of a mountain

RUBRIC FOR GRAPHS: GEARING UP!, CREVASSE CRISIS!, AND SLIDING DOWN!

	EXPERT (4)	COMPETENT (3)	BEGINNER (2)	NOVICE (1)
Labeling	☐ Clearly and appropriately labels the *x*- and *y*-axes ☐ Labels graph with a title that correctly identifies the data being represented ☐ Correctly identifies independent and dependent variables	☐ Appropriately labels the *x*- and *y*-axes ☐ Labels graph with a title that correctly identifies the data being represented ☐ Correctly identifies independent and dependent variables	☐ Leaves out one of the axis labels or inappropriately labels one axis (e.g., forgets to include units) ☐ Leaves out title or uses a title that partially identifies the data being represented ☐ Does not correctly identify independent and dependent variables	☐ Leaves out both axis labels or inappropriately labels both axes ☐ Leaves out title or uses title that does not identify the data being represented ☐ Does not correctly identify independent and dependent variables
Scales and intervals	☐ Uses scales and intervals for *x*- and *y*-axes that show the entire range of data; date is well spread out ☐ Shows equal intervals	☐ Uses scales and intervals for *x*- and *y*-axes that show the entire range of data, but data may not be well spread out ☐ Shows mostly equal intervals (one or two minor errors)	☐ Scales on *x*- and *y*-axes may not reflect the range of data needed, or intervals may not be appropriate to precisely show data. ☐ Shows some equal intervals (a few major errors on one or both axes)	☐ Missing scale on one or more axes, scales on *x*- and *y*-axes do not reflect the range of data needed, or intervals are not appropriate to show the data. ☐ Shows unequal intervals (e.g., uses data values as intervals)
Data representation	☐ Graph accurately reflects data in the data table.	☐ Graph mostly reflects data in the data table (one or two minor errors).	☐ Graph reflects some of the data in the data table (many minor errors or a few major errors).	☐ Graph reflects few to none of the data in the data table (major errors).
Type of graph	☐ Type of graph is appropriate for the kind of data.	☐ Type of graph is appropriate for the kind of data.	☐ Type of graph may not be appropriate for the kind of data.	☐ Type of graph is not suitable to represent the kind of data.

STUDENT PAGE

RUBRIC FOR ENGINEERING DRAWINGS (STEP 4): GEARING UP!

	EXPERT (4)	COMPETENT (3)	BEGINNER (2)	NOVICE (1)
Quality of idea	☐ Selects a design that addresses the problem ☐ Uses materials efficiently and with purpose ☐ Chooses design as a team through thoughtful deliberation ☐ Is able to express rationale for each part of the design	☐ Selects a design that addresses most of the problem ☐ Uses material with some efficiency and with purpose ☐ Chooses design as a team after some deliberation ☐ Is able to express rationale for most parts of the design	☐ Selects a design that somewhat addresses the problem ☐ Uses materials with some purpose ☐ Chooses design as a team after a little discussion ☐ Is able to express rationale for some parts of the design	☐ Selects a design that does not address the problem ☐ Does not use materials with purpose ☐ Chooses design hastily without much discussion ☐ Is not able to express rationale for design
Communication	☐ Labels all the materials used in the design ☐ Shows how all materials are layered	☐ Labels most of the materials used in the design ☐ Shows how most of the materials are layered	☐ Labels some of the materials used in the design ☐ Shows how some of the materials are layered	☐ Does not label materials used in the design ☐ Does not show how materials are layered

STUDENT PAGE

RUBRIC FOR ENGINEERING DRAWINGS (STEP 4): CREVASSE CRISIS! AND SLIDING DOWN!

	EXPERT (4)	COMPETENT (3)	BEGINNER (2)	NOVICE (1)
Quality of idea	☐ Selects a design that addresses the problem ☐ Uses materials efficiently and with purpose ☐ Chooses design as a team through thoughtful deliberation ☐ Is able to express rationale for each part of the design	☐ Selects a design that addresses most of the problem ☐ Uses material with some efficiency and with purpose ☐ Chooses design as a team after some deliberation ☐ Is able to express rationale for most parts of the design	☐ Selects a design that somewhat addresses the problem ☐ Uses materials with some purpose ☐ Chooses design as a team after a little discussion ☐ Is able to express rationale for some parts of the design	☐ Selects a design that does not address the problem ☐ Does not use materials with purpose ☐ Chooses design hastily without much discussion ☐ Is not able to express rationale for design
Communication	☐ Labels all dimensions ☐ Uses appropriate units ☐ Labels all the materials used in the design ☐ Shows how all materials are joined	☐ Labels most dimensions ☐ Uses appropriate units ☐ Labels most of the materials used in the design ☐ Shows how most of the materials are joined	☐ Labels some dimensions ☐ Uses somewhat appropriate units ☐ Labels some of the materials used in the design ☐ Shows how some of the materials are joined	☐ Does not label dimensions ☐ Units are not included or are inappropriate. ☐ Does not label materials used in the design ☐ Does not show how materials are joined

Name
STUDENT PAGE

RUBRIC FOR PROTOTYPE/MODEL (STEP 5): GEARING UP!, CREVASSE CRISIS!, AND SLIDING DOWN!

	EXPERT (4)	COMPETENT (3)	BEGINNER (2)	NOVICE (1)
Completeness	☐ Builds a model that meets all criteria and constraints ☐ Follows the design sketch ☐ Follows cleanup procedures	☐ Builds a model that addresses most of the criteria and constraints ☐ Follows most of the design sketch ☐ Follows cleanup procedures	☐ Builds a model that addresses some of the criteria and constraints ☐ Follows some of the design sketch ☐ Partially follows cleanup procedures	☐ Builds an incomplete model ☐ Does not follow the design sketch ☐ Does not follow cleanup procedures
Craftsmanship	☐ Takes care in constructing model; is adept with tools and resources, making continual adjustments to "tweak" the model/prototype ☐ Demonstrates persistence with minor problems	☐ Uses tools and resources with little or no guidance ☐ Refines model to enhance appearance and capabilities	☐ Uses tools and resources with some guidance; may have difficulty selecting the appropriate resource ☐ Refines work, but may prefer to leave model as first produced	☐ Needs guidance in order to use resources safely and appropriately ☐ Model/prototype is crude, with little or no refinements made

RUBRIC FOR TEST, COMMUNICATE, AND REDESIGN STEPS: GEARING UP!, CREVASSE CRISIS!, AND SLIDING DOWN!

	EXPERT (4)	COMPETENT (3)	BEGINNER (2)	NOVICE (1)
Completeness	☐ Carefully follows the testing procedures and documents all testing results	☐ Follows the testing procedures and documents most of the testing results	☐ Follows some of the testing procedures and documents some of the testing results	☐ Does not follow the testing procedures and does not document the testing results
Model performance	☐ The model fully meets the design constraints and criteria.	☐ The model meets most of the design constraints and criteria.	☐ The model meets some design constraints and criteria completely but ignores others.	☐ The model fails to meet design criteria and constraints.
Quality of reflection	☐ Specific improvement ideas are generated and documented.	☐ Some general improvement ideas are generated and documented.	☐ The need for improvements is recognized and some ideas are generated but documentation is not complete.	☐ Little interest is taken in improving the prototype or model, despite problems detected during testing. There is no evidence of inclination or ability to generate refinement solutions.

STUDENT PARTICIPATION RUBRIC

	0 POINTS	1 POINT	2 POINTS	3 POINTS	SCORE
CONTENT CONTRIBUTION					
Sharing information	Discussed very little information related to the topic	Discussed some basic information; most related to the topic	Discussed a great deal of information; all related to the topic	Discussed a great deal of information showing in-depth analysis and thinking skills	
Creativity	Did not contribute any new ideas	Contributed some new ideas	Contributed many new ideas	Contributed a great deal of new ideas	
RESPONSIBILITY					
Completion of assigned duties	Did not perform any assigned duties	Performed very few assigned duties	Performed nearly all assigned duties at the level of expectation	Performed all assigned duties; did extra duties	
Attendance	Was never present or was always a negative influence when present	Attended some group meetings; absence(s) hurt the group's progress	Attended most group meetings; absence(s) did not affect group's progress or made up work	Attended all focus group meetings	
Staying on task	Not productive during group meetings; often distracted the team	Productive some of the time; needed reminders to stay on task	Productive most of the time; rarely needed reminders to stay on task	Used all of the focus group time effectively; productive at all times	
TEAMWORK					
Cooperating with teammates	Was rarely talking or always talking; usually argued with teammates	Usually did most of the talking; rarely allowed others to speak; sometimes argued	Listened, but sometimes talked too much; rarely argued	Listened and spoke a fair amount; never argued with teammates	
Making fair decisions	Always needed to have things his or her way; easily upset	Usually wanted to have things his or her way or often sided with friends instead of considering all views	Usually considered all views	Always helped team to reach a fair decision	
Leadership	Never took lead; needed to be assigned duties	Took a lead at least once; volunteered for duty	Took a lead more than once; volunteered for duties and helped others	Played essential role in organizing the group; frequently took lead; always helped others	/ 24

Name: _____ Teacher: _____ Total score: _____

STUDENT WORK SAMPLE 1

GEARING UP! STEP 4: CHOOSE THE BEST SOLUTION

Use the rubric provided by your teacher to assess the following student work sample. Write a brief explanation for the grade you assign and how the work can be improved.

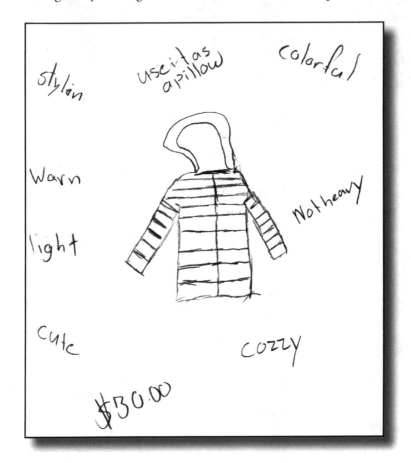

1. Grade: _____

2. Reasons for grade: _____

    _____

    _____

    _____

3. How work can be improved: _____

    _____

    _____

    _____

STUDENT WORK SAMPLE 2

GEARING UP! STEP 4: CHOOSE THE BEST SOLUTION

Use the rubric provided by your teacher to assess the following student work sample. Write a brief explanation for the grade you assign and how the work can be improved.

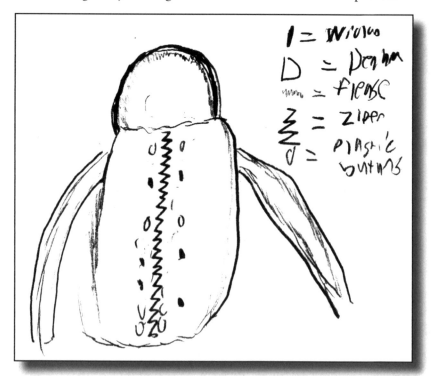

1. Grade: _____

2. Reasons for grade: _____

   _____

   _____

   _____

3. How work can be improved: _____

   _____

   _____

   _____

STUDENT WORK SAMPLE 3

GEARING UP! STEP 4: CHOOSE THE BEST SOLUTION

Use the rubric provided by your teacher to assess the following student work sample. Write a brief explanation for the grade you assign and how the work can be improved.

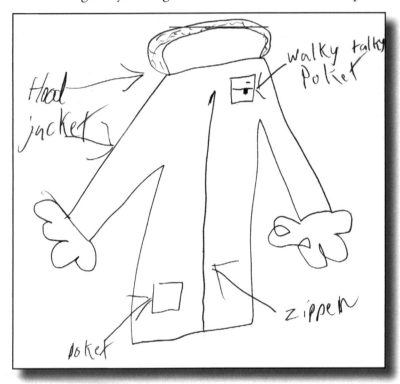

1. Grade: _____

2. Reasons for grade: _____

    _____

    _____

    _____

3. How work can be improved: _____

    _____

    _____

    _____

STUDENT WORK SAMPLE 4

GEARING UP! STEP 4: CHOOSE THE BEST SOLUTION

Use the rubric provided by your teacher to assess the following student work sample. Write a brief explanation for the grade you assign and how the work can be improved.

1. Grade: _____

2. Reasons for grade: _____

   _____

   _____

   _____

3. How work can be improved: _____

   _____

   _____

   _____

STUDENT WORK SAMPLE 5

GEARING UP! STEP 4: CHOOSE THE BEST SOLUTION

Use the rubric provided by your teacher to assess the following student work sample. Write a brief explanation for the grade you assign and how the work can be improved.

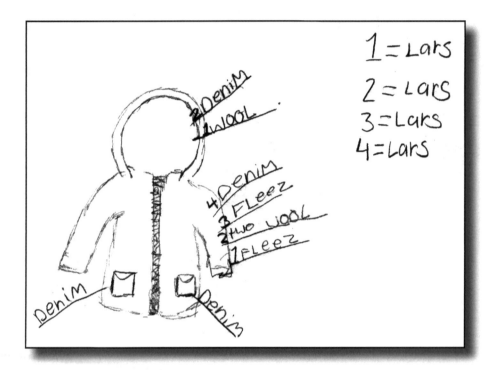

1. Grade: _____

2. Reasons for grade: _____

    _____

    _____

    _____

3. How work can be improved: _____

    _____

    _____

    _____

STUDENT WORK SAMPLE 6

SLIDING DOWN! STEP 4: CHOOSE THE BEST SOLUTION

Use the rubric provided by your teacher to assess the following student work sample. Write a brief explanation for the grade you assign and how the work can be improved.

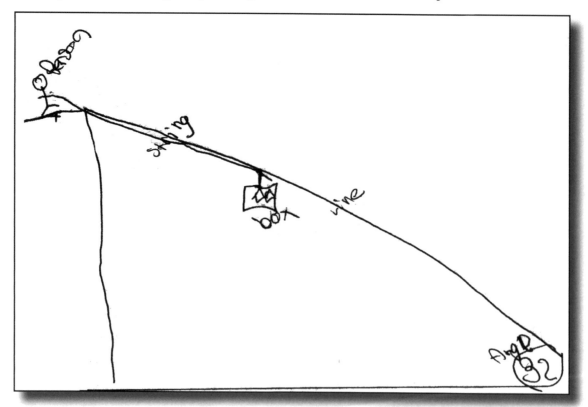

1. Grade: _____

2. Reasons for grade: _____

    _____

    _____

    _____

3. How work can be improved: _____

    _____

    _____

    _____

STUDENT WORK SAMPLE 7

SLIDING DOWN! STEP 4: CHOOSE THE BEST SOLUTION

Use the rubric provided by your teacher to assess the following student work sample. Write a brief explanation for the grade you assign and how the work can be improved.

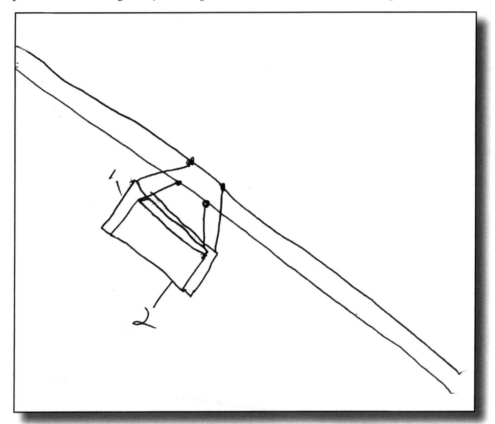

1. Grade: _____

2. Reasons for grade: _____

    _____

    _____

    _____

3. How work can be improved: _____

    _____

    _____

    _____

STUDENT WORK SAMPLE 8

SLIDING DOWN! STEP 4: CHOOSE THE BEST SOLUTION

Use the rubric provided by your teacher to assess the following student work sample. Write a brief explanation for the grade you assign and how the work can be improved.

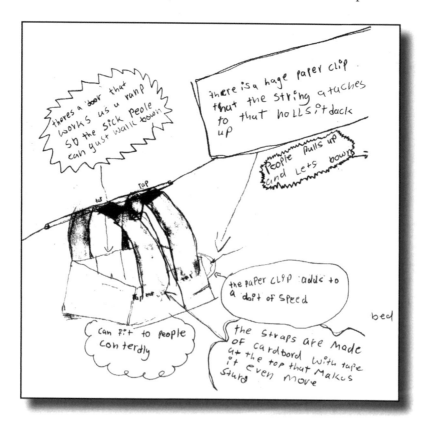

1.   Grade: _____

2.   Reasons for grade: _____

     _____

     _____

     _____

3.   How work can be improved: _____

     _____

     _____

     _____

STUDENT WORK SAMPLE 9

SLIDING DOWN! STEP 4: CHOOSE THE BEST SOLUTION

Use the rubric provided by your teacher to assess the following student work sample. Write a brief explanation for the grade you assign and how the work can be improved.

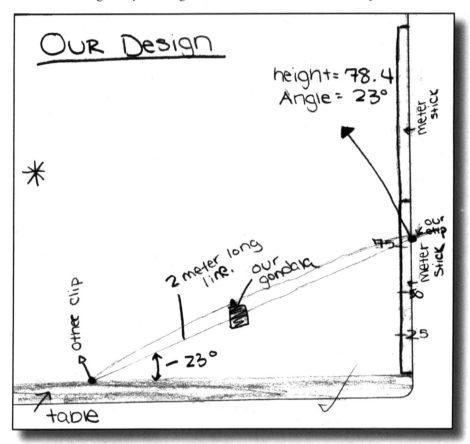

1. Grade: _____

2. Reasons for grade: _____

    _____

    _____

    _____

3. How work can be improved: _____

    _____

    _____

    _____

Answer Key

LINE GRAPH ACTIVITY

Exercise 1

1. Mileage; it is an "input."
2. Value; it is an "output."
3. *x* mileage (independent variable)
4. *y* value (dependent variable)
5. a. 0–120,000; 120,000
   b. 25
   c. 120,000, 25; data range is greater than number of boxes so 120,000 ÷ 25 = 4,800
   d. Every box is worth 4,800.
   e. *x*-axis should be labeled "Mileage" with appropriate units.

6–7. See sample graph; answers will vary.

Sample graph:

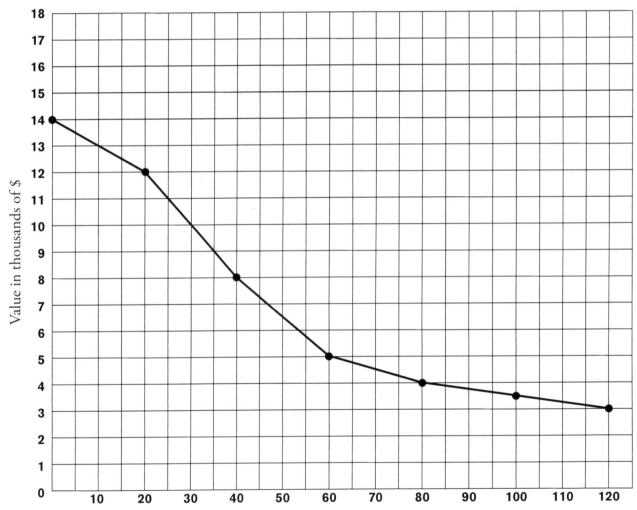

Relationship Between Truck Mileage and Value

8. Answers will vary.
9. $3,500
10. 20,000–40,000 kilometers
11. $6,500
12. $2,500, by extending the line or estimating the next decrease based on the previous decrease

Sample graph:

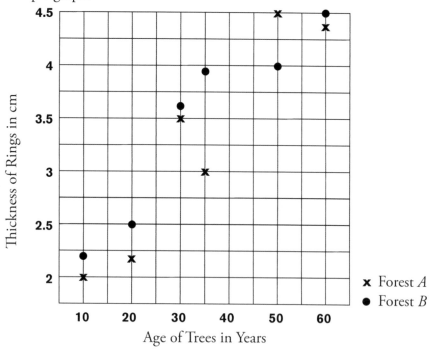

Age of Trees in Years

Exercise 2
1. 2.5 cm
2. 30 or 60 years old
3. 5.0 cm, based on the general trend (up). However, this is not guaranteed due to the unpredictability of climate and weather.
4. Rainfall, for example, has been more irregular in Forest *A* than in Forest *B*. The pattern of growth is more erratic in Forest *A* than Forest *B*.

PAGE 14

USING A SCALE ACTIVITY
1. 4
2. 36 centimeters
3. .3 meters
4. 4 m
5. 10 m
6. 2 cm

PAGES 26–27

INTRODUCING THE ENGINEERING DESIGN PROCESS (EDP)
1. Answers will vary.
2. Step 1: Define the problem.
   Step 2: Research the problem.
   Step 3: Brainstorm possible solutions.
   Step 4: Choose the best solution.
   Step 5: Build a model or prototype.
   Step 6: Test your solution.
   Step 7: Communicate your solution.
   Step 8: Redesign as needed.

3.   a. Step 2: Research
     b. Step 5: Build
     c. Step 7: Communicate
     d. Step 1: Define
     e. Step 4: Choose
     f. Step 8: Redesign
     g. Step 6: Test
     h. Step 3: Brainstorm

PAGE 36

DESIGN CHALLENGE 1: GEARING UP!
2. Research the Problem
Research Phase 1: What Do We Know?
1. 8.2°C
2. 1.3°C
3. Temperature decreases over time, but the rate of change in the temperature is also decreasing, so temperature decreases more slowly over time.
4. a. A reasonable answer would be around –20°C.
   b. The change in temperature is decreasing over time: 45–50 seconds (D_{temp} = –1.9°), 50–55 seconds (D_{temp} = –1.5°), 55–60 seconds (D_{temp} = –1.3°). Therefore, at 60–65 seconds, the temperature should decrease, but by slightly less than 1.3°. At 65 seconds, the temperature should be less than –19.1°C, but greater than –20.4°C.
5. a. –26°C
   b. The temperature of the thermometer will approach equilibrium with the environment, set at –26°C.
6. No. After 30 seconds, the temperature inside the cotton was –6.8°C. The temperature dropped below 18°C after just 5 seconds.
7. Any graph with a value remaining above 18°C for 30 seconds would be correct.

PAGE 48

Research Phase 2: Learning About Materials
Experiment 2: What Is the Effect of Layering One Material?
1. Graphs will vary but should reflect the data recorded in Table 1.3.
2. In general, as the number of layers increases, the temperature inside the material after 30 seconds increases.
3–4. Answers will vary.

DESIGN CHALLENGE 2: CREVASSE CRISIS!

1. Define the Problem: Figuring Out the Scale

1. 30 cm : 5 m

   Therefore, proportionally, 6 cm : 1 m.

   So 6 cm represents 1 m.

2a. $\dfrac{6 \text{ cm}}{1 \text{ m}} = \dfrac{x}{2 \text{ m}}$

   $x = 12$ cm

   The scale model will have a 12-cm-wide crevasse.

2b. $\dfrac{6 \text{ cm}}{1 \text{ m}} = \dfrac{x}{0.2 \text{ m}}$

   $x = 1.2$ cm

   The maximum sag allowed in the scale model is 1.2 cm.

2c. $\dfrac{6 \text{ cm}}{1 \text{ m}} = \dfrac{x}{0.4 \text{ m}}$

   $x = 2.4$ cm

   The scale model bridge must be at least 2.4 cm wide.

PAGES 78–79

2. Research the Problem

Research Phase 1: Bending Under a Weight

1. Answers will vary. Possible answers: the strength of the material itself; the length of the bridge; the width of the bridge; the thickness of the bridge; the shape and structure of the bridge.

2. Answers will vary.

3. The craft stick should be easy to snap in half. The flat part of the stick is thin and not very strong.

4. It should be more difficult to break the craft stick along the thickest part.

5. Answers will vary.

PAGES 82–83

Research Phase 2: How Does Width Affect Bending?

1. Answers will vary.

2. x-axis: Width of Ladder (cm); y-axis: Amount of Deflection (cm)

3. In general, as width increases, the ladder deflects/sags less.

4–5. Answers will vary.

PAGES 86–87

Research Phase 3: How Does the Thickness Affect Bending?

1. Answers will vary.

2. x-axis: Thickness (number of ladders/cm); y-axis: Amount of Deflection (cm)

3. In general, as thickness increases, the bridge deflects/sags less.

4. Students should find that thickness affects strength more than width. Specifically, strength grows linearly with width but grows cubically with thickness.

5. The wide side has some width, but very little thickness. The thin side has little width, but is much thicker. The thin side is thicker than the wide side by the same factor as the wide side is wider than the thin side. Since thickness contributes more to strength than width does, the thin side should be able to withstand more force.

3. Brainstorm Possible Solutions
  1. 1.2 cm; 2.4 cm
  2. Sketches will vary.

DESIGN CHALLENGE 3: SLIDING DOWN!

Problem Solving

Tank 3 + 4 = 70 min $135; this is the best choice.
Tank 2 + 5 = 70 min $145
Tank 2 + 4 = 60 min $120

2. Research the Problem
Research Phase 1: Investigating Angles to Find an Adequate Speed
Table below shows approximate values.

HEIGHT OF BEGINNING OF ZIP-LINE (M)	0.4	0.6	0.8	1.0	1.2	1.4
ANGLE (DEGREES)	12	17	24	30	37	45
TIME (SECONDS)	Doesn't move	3.19	1.47	1.17	0.83	0.67

  1. Answers will vary. Taking accurate time measurements is difficult because the Timer may not react fast enough to stop the stopwatch at the exact moment the straw stops moving.
2–3. Answers will vary.
  4. *x*-axis: Angle (degrees); *y*-axis: Speed (m/s)
  5. In general, as angle increases, speed increases, because the straw is traveling down a steeper slope.
6–8. Answers will vary.

Research Phase 2: How Can You Make the Transporter Safe?
There are many possible factors. For a platform that runs above the zip-line, factors include number of zip-lines, distances between zip-lines, and the angle of the platform on which the passengers sit. For a transporter that hangs below the zip-line, factors include the number of ropes or connectors used to hang the transporter and the number of zip-lines.

Student Work Sample 1
  1. 1
  2. The drawing doesn't address the problem criteria or constraints. It doesn't show how materials are used or layered. There is no labeling of materials or optional parts. It is unclear how cost was determined.
  3. Label materials used and optional items (for example, hood material). Show thickness measurement. Show how layers are ordered from the inside out and label the materials.

Student Work Sample 2

1. 2
2. The drawing includes a key. You can identify where the zipper, buttons, and some of the layer materials are located on the coat, but it is unclear where denim material is used. It is also unclear how many layers of each material are used. The drawing somewhat shows the order of layers but could be more clear.
3. Show number of layers of each material used in coat. Show measurement for the thickness of layers.

Student Work Sample 3

1. 2
2. Optional items are clearly labeled, but layers are not shown or labeled. Drawing doesn't show number of layers of each material used.
3. Show and label materials for layers, number of layers of each material, and location of layers. Identify material used for hood and pockets. Show measurement for thickness of layers.

Student Work Sample 4

1. 3
2. Optional items are shown and labeled, including the hood and pockets. Drawing shows cross section of materials, but the number and position of layers are not clearly indicated.
3. Show order of layers more clearly (label which is the outer layer and which is the innermost layer). Clearly indicate number of layers of each material. Show measurement for thickness of layers.

Student Work Sample 5

1. 3
2. Drawing shows materials used and their locations on the coat. Drawing shows how the materials are layered, but it is unclear which are the outer and inner layers. It is unclear whether there is a zipper on the coat because it's not labeled.
3. Show order and number of layers more clearly. Label zipper if there is one. Show measurement of thickness of layers.

Student Work Sample 6

1. 2
2. Drawing shows all major parts of the design solution, but the only dimension labeled is the angle.
3. Label more dimensions, such as the length of the string, the height of the starting location, and the dimensions for the transporter.

Student Work Sample 7

1. 1
2. Drawing shows a close-up of the transporter and includes some numbers that suggest measurement, but units are not included. Drawing doesn't show any other part of the design needed to address the problem's criteria and constraints.
3. Draw the rest of the design and include labels to show amount of string needed, angle, starting height, and return mechanism. Also include dimensions and units for all of the above.

Student Work Sample 8

1. 2
2. Drawing includes a nice artistic rendering of the transporter and a good explanation describing the rationale for its design. Materials are labeled. However, the drawing is missing the rest of the design to fully meet the problem's criteria and constraints.
3. Draw the rest of the design solution showing angle, return mechanism, length of string, and starting height. Label with dimensions and units.

Student Work Sample 9

1. 4
2. Drawing shows a complete solution that includes the starting height, angle, labeled materials, and dimensions.
3. Show a closer view of the transporter (with dimensions) and explain how the return mechanism works.